Lecture Notes in Mathematics

Edited by A. Dold and B. Eckmann

434

Philip Brenner
Vidar Thomée
Lars B. Wahlbin

Besov Spaces and Applications to Difference Methods for Initial Value Problems

Springer-Verlag
Berlin · Heidelberg · New York 1975

Dr. Philip Brenner
Prof. Vidar Thomée
Department of Mathematics
Chalmers University of Technology
and University of Göteborg, Fack
S–402 20 Göteborg 5/Sweden

Prof. Lars B. Wahlbin
Department of Mathematics
Cornell University
White Hall
Ithaca, NY 14850/USA

Library of Congress Cataloging in Publication Data

Brenner, Philip, 1941–
  Besov spaces and applications to difference methods
for initial value problems.

  (Lecture notes in mathematics ; 434)
  Includes bibliographies and index.
  1. Differential equations, Partial. 2. Initial
value problems. 3. Besov spaces. I. Thomée,
Vidar, 1933–   joint author. II. Wahlbin, Lars
Bertil, 1945–   joint author. III. Title.
IV. Series.
QA3.L28 no. 434 ₍QA377₎ 510'.8s ₍515'.353₎ 74-32455

AMS Subject Classifications (1970): 35 E 15, 35 L 45, 42 A 18, 46 E 35, 65 M 10, 65 M 15

ISBN 3-540-07130-X Springer-Verlag Berlin · Heidelberg · New York
ISBN 0-387-07130-X Springer-Verlag New York · Heidelberg · Berlin

Offsetdruck: Julius Beltz, Hemsbach/Bergstr.

PREFACE

The purpose of these notes is to present certain Fourier techniques for analyzing finite difference approximations to initial value problems for linear partial differential equations with constant coefficients. In particular, we shall be concerned with stability and convergence estimates in the L_p norm of such approximations; the main theme is to determine the degree of approximation of different methods and the precise dependence of this degree upon the smoothness of the initial data as measured in L_p. In L_2 the analysis generally depends on Parseval's relation and is simple; it is to overcome the difficulties present in order to obtain estimates in the maximum-norm, or more generally in L_p with $p \neq 2$, which is the aim of this study.

The main tools which we shall use are some simple results on Fourier multipliers based on inequalities by Carlson and Beurling and by van der Corput. Many results are expressed in terms of norms in Besov spaces $B_p^{s,q}$ where s essentially describes the degree of smoothness with respect to L_p.

The first two chapters contain the prerequisits on Fourier multipliers and on Besov spaces, respectively, needed for our applications. The purpose of these two chapters is only to make these notes self-contained and not to give an extensive treatment of their topics. Chapters 3 through 6 then form the main part of the notes. In Chapter 3 we present preliminary material on initial value problems and finite difference schemes for such problems. In particular, the concepts of well-posedness in L_p of an initial value problem and stability in L_p and accuracy of a finite difference approximation are defined and expressed in terms of Fourier transforms, and estimates which are based on simple analysis in L_2 are derived. The remaining chapters are then devoted to the more refined results in L_p with $p \neq 2$ for the heat equation, first order hyperbolic equations and the Schrödinger equation, respectively.

Except for some results in Chapter 6, the material in these notes can be found in papers published by the authors and others. Rather than striving for generality we have chosen, for the purpose of making the techniques transparent, to treat only simple cases.

The results and formulae are numbered by chapter, section, and order within each section so that, for instance, Theorem 1.2.3 means the third theorem of Chapter 1, Section 2 (or Section 1.2). For reference within a chapter the first number is dropped so that the above theorem within Chapter 1 is referred to as Theorem 2.3. The references to the literature are listed at the end of each chapter.

Throughout these notes, C and c will denote large and small positive constants, respectively, not necessarily the same at different occurrences.

The work of the latter two authors has been supported in part by the National Science Foundation, USA.

Göteborg, Sweden and Ithaca, N.Y., USA in September 1974

TABLE OF CONTENTS

CHAPTER 1. FOURIER MULTIPLIERS ON L_p.

In this chapter we develop the theory of Fourier multipliers on L_p to the extent needed for the applications in later chapters. Since our applications are quantitative rather than qualitative, we shall define the L_p multiplier norm $M_p(a)$ for smooth a only, and our efforts will then be to describe some techniques to estimate this norm. In Section 1 we introduce the necessary definitions and in Section 2 we then collect a number of basic properties of the multiplier norms. In Section 3 we derive an inequality for $M_\infty(a)$ by Carlson and Beurling which will be one of our main tools later. In Section 4 we reduce the problem of estimating periodic multipliers to the corresponding problem for multipliers with compact support, and in Section 5, finally, we prove a lemma by van der Corput and some consequences relevant to the present context.

1.1. Preliminaries and definition.

For $x = (x_1,\ldots,x_d) \in R^d$ and $\xi = (\xi_1,\ldots,\xi_d) \in R^d$, let $\langle x,\xi\rangle = x_1\xi_1+\ldots+x_d\xi_d$ and $|x| = \langle x,x\rangle^{1/2}$. We shall use the Fourier transform normalized so that for functions $u \in L_1$,

$$\mathcal{F}u(\xi) \equiv \hat{u}(\xi) = \int e^{-i\langle x,\xi\rangle}u(x)dx .$$

Its inverse is then formally

$$\mathcal{F}^{-1}v(x) \equiv \check{v}(x) = (2\pi)^{-d}\int e^{i\langle x,\xi\rangle}v(\xi)d\xi ,$$

and the Fourier inversion formula $\mathcal{F}^{-1}\hat{u} = u$ holds if u and $\hat{u}$ both belong to L_1. Parseval's formula now reads

$$\int u\bar{v}dx = (2\pi)^{-d}\int \hat{u}\bar{\hat{v}}d\xi .$$

(Unless specified to the contrary all functions considered will be complex-valued.)

For $\alpha = (\alpha_1,\ldots,\alpha_d)$ a non-negative multi-integer we define

$$\xi^\alpha = \xi_1^{\alpha_1} \ldots \xi_d^{\alpha_d}, \quad D^\alpha = \left(\frac{\partial}{\partial x_1}\right)^{\alpha_1} \ldots \left(\frac{\partial}{\partial x_d}\right)^{\alpha_d},$$

and have then

$$D^\alpha u(x) = \mathcal{F}^{-1}((i\xi)^\alpha \hat{u})(x).$$

Further, for $y \in R^d$,

$$u(x+y) = \mathcal{F}^{-1}(e^{i\langle y,\xi\rangle}\hat{u})(x), \tag{1.1}$$

and

$$(u * v)(x) \equiv \int u(x-y)v(y)dy = \mathcal{F}^{-1}(\hat{u}\hat{v})(x).$$

Let $\hat{C}_0^\infty = \hat{C}_0^\infty(R^d)$ denote the class of functions which have Fourier transforms in $C_0^\infty = C_0^\infty(R^d)$, the set of functions $v \in C^\infty(R^d)$ with $\operatorname{supp}(v)$ (the support of v) compact. In this chapter we shall mainly work with the Fourier transform and its inverse acting on functions in $\hat{C}_0^\infty$ and C_0^∞, respectively, but in later chapters the Fourier transform (and differentiation) will be applied more generally to elements in the space S' of tempered distributions, in particular to functions in L_p, $1 \le p \le \infty$.

The norms in the spaces L_p, $1 \le p \le \infty$, are given as usual by

$$\|u\|_p = \begin{cases} \left(\int |u(x)|^p dx\right)^{1/p} & \text{for } 1 \le p < \infty, \\ \operatorname{ess\,sup}_{x\in R^d} |u(x)| & \text{for } p = \infty. \end{cases}$$

We shall denote by W_p the closure in the L_p norm of $\hat{C}_0^\infty$ (or C_0^∞, or S, the class of functions which together with all their derivatives tend to zero faster than any negative power of $|x|$, as $|x|$ tends to infinity). For $1 \le p < \infty$ we have $W_p = L_p$ whereas W_∞ is the space of continuous functions which vanish at infinity and is properly contained in L_∞.

Let now for $a \in C^\infty(R^d)$ the operator A from $\hat{C}_0^\infty$ into itself be defined by

$$Au = \mathcal{F}^{-1}(a\hat{u}). \tag{1.2}$$

It is easily seen from (1.1) that A is translation invariant so that for any $y \in R^d$,

$$A(u(\cdot - y))(x) = Au(x-y).$$

If A is a given translation invariant operator of the form (1.2), the function a is referred to as its symbol and is often denoted by $\hat{A}$.

In Chapter 2 ff. we shall mainly work with functions a in the class of slowly increasing functions, i.e. functions which together with all their derivatives have at most polynomial growth. In this case we may consider A defined by (1.2) as a continuous operator in S'.

We say that a is a Fourier multiplier on L_p or that $a \in M_p = M_p(R^d)$ if

$$M_p(a) \equiv \sup\{\|Au\|_p : u \in \hat{C}_0^\infty,\ \|u\|_p \leq 1\} < \infty .$$

The operator A on $\hat{C}_0^\infty$ defined by (1.2) may then be extended by completion from $\hat{C}_0^\infty$ to W_p. Since L_p is continuously embedded in S' this extension is consistent with the distribution interpretation of (1.2) when a is slowly increasing. We shall see later (Theorem 2.3) that in the case $p = \infty$ we can similarly extend A to a bounded linear operator not only on W_∞ but on L_∞.

Occasionally we shall use the notation $M_p^{(d)}$ and $M_p^{(d)}(\cdot)$ for the multipliers and their norms in order to emphasize the dimension of the underlying space R^d. Notice that in this presentation multipliers are always C^∞ functions.

1.2. Basic properties.

We first show that M_p is symmetric with respect to conjugate indices.

Theorem 2.1. Let $1 \leq p,\ p' \leq \infty$, $1/p + 1/p' = 1$. Then $M_p = M_{p'}$ and for $a \in C^\infty$,

$$M_p(a) = M_{p'}(a).$$

Proof. Let $u_-(x) = u(-x)$. Using (1.2) we have by Parseval's formula, Hölder's inequality, and the fact that $(\hat{u})_- = \widehat{(u_-)}$, that for $u,v \in \hat{C}_0^\infty$,

$$\left|\int Av \cdot u\,dx\right| = \left|\int \mathscr{F}^{-1}(a\hat{v})\overline{\mathscr{F}^{-1}\hat{\bar{u}}}\,dx\right| = (2\pi)^{-d}\left|\int a\hat{v}(\hat{u})_-\,d\xi\right|$$

$$= \left|\int \mathscr{F}^{-1}(a\widehat{(u_-)})v_-\,dx\right| \leq \|Au_-\|_p\|v\|_{p'} \leq M_p(a)\|u\|_p\|v\|_{p'}.$$

Hence by the converse of Hölder's inequality,

$$\|Av\|_{p'} \leq M_p(a)\|v\|_{p'},$$

that is, $M_{p'}(a) \leq M_p(a)$. Reversing the roles of p and p' we obtain the desired result.

We next give characterizations of M_2 and M_∞ (= M_1 by Theorem 2.1).

<u>Theorem 2.2</u>. M_2 consists of the uniformly bounded functions in C^∞ and for $a \in M_2$,

$$M_2(a) = \|a\|_\infty.$$

<u>Proof</u>. Assume first that a is bounded. Then with A defined by (1.2) we have for $u \in \hat{C}_0^\infty$,

$$\|Au\|_2 = (2\pi)^{-d/2}\|a\hat{u}\|_2 \leq (2\pi)^{-d/2}\|a\|_\infty\|\hat{u}\|_2 = \|a\|_\infty\|u\|_2,$$

and hence $a \in M_2$ and

$$M_2(a) \leq \|a\|_\infty. \tag{2.1}$$

Conversely, let $a \in M_2$ and let $\xi_0 \in R^d$ and $\varepsilon > 0$ be arbitrary. Then there exists a sphere B with center at ξ_0 such that

$$|a(\xi)| \geq |a(\xi_0)|(1-\varepsilon) \quad \text{for } \xi \in B. \tag{2.2}$$

Let $u_0 \in \hat{C}_0^\infty$, $u_0 \not\equiv 0$, be such that $\operatorname{supp}(\hat{u}_0) \subset B$. Then Parseval's formula and (2.2) give

$$\|Au_0\|_2 = (2\pi)^{-d/2}\|a\hat{u}_0\|_2 \geq (2\pi)^{-d/2}|a(\xi_0)|(1-\varepsilon)\|\hat{u}_0\|_2 = |a(\xi_0)|(1-\varepsilon)\|u_0\|_2.$$

Since ξ_0 and ε are arbitrary, we conclude that a is bounded and that

$$\|a\|_\infty \leq M_2(a).$$

Together with (2.1) this completes the proof of the theorem.

For the characterization of M_∞, let B denote the set of bounded complex regular measures on R^d with the total variation norm $V(\cdot)$. Recall that for $\mu(x) = f(x)dx$ with $f \in L_1$, $V(\mu) = \|f\|_1$, and that the convolution between a function and a measure is defined by

$$u * \mu(x) = \int u(x-y)d\mu(y).$$

Our next result shows that the elements of M_∞ (or M_1) are Fourier transforms of measures in B.

<u>Theorem 2.3</u>. Let $a \in M_\infty$. Then there exists $\mu \in B$ such that

$$a(\xi) = \int e^{-i\langle x,\xi\rangle} d\mu(x), \tag{2.3}$$

$$M_\infty(a) = V(\mu), \tag{2.4}$$

$$Au \equiv \mathcal{F}^{-1}(a\hat{u}) = u * \mu \quad \text{for} \quad u \in \hat{C}_0^\infty. \tag{2.5}$$

Conversely, let $a \in C^\infty$ and assume that (2.3) holds with $\mu \in B$. Then $a \in M_\infty$ and (2.4), (2.5) hold true.

<u>Proof</u>. Assume first that $a \in M_\infty$. We have for the operator A,

$$|Au(0)| \leq M_\infty(a)\|u\|_\infty.$$

Hence the linear form $u \rightsquigarrow Au(0)$ may be extended to a bounded linear functional on W_∞. By the Riesz representation theorem there exists a measure μ in B such that

$$Au(0) = \int u(-y)d\mu(y) \quad \text{for} \quad u \in W_\infty.$$

Since the operator A is translation invariant it follows that

$$Au(x) = A(u(\cdot+x))(0) = \int u(x-y)d\mu(y) = u*\mu(x),$$

which proves (2.5).

By the Riesz representation theorem we also have for each fixed x that

$$\sup_{u \in \hat{C}_0^\infty} \frac{|Au(x)|}{\|u\|_\infty} = V(\mu),$$

and hence the norm equality (2.4) follows easily.

It remains to prove (2.3). Fourier transformation of (2.5) gives for $u \in \hat{C}_0^\infty$,

$$a\hat{u} = \mathcal{F}(u*\mu). \tag{2.6}$$

The right hand side may be calculated as follows:

$$\mathcal{F}(u*\mu)(\xi) = \int e^{-i<x,\xi>} \int u(x-y)d\mu(y)dx$$

$$= \int \left(\int e^{-i<x,\xi>}u(x-y)dx\right)d\mu(y) = \hat{u}(\xi) \int e^{-i<y,\xi>}d\mu(y).$$

Here the change in the order of integration is justified by the Fubini-Tonelli theorem since

$$\int \left(\int |e^{-i<x,\xi>}u(x-y)|dx\right)d|\mu|(y) \leq \|u\|_1 V(\mu).$$

Hence it follows from (2.6) that

$$a(\xi)\hat{u}(\xi) = \hat{u}(\xi) \int e^{-i<y,\xi>}d\mu(y),$$

which proves (2.3).

For the converse we find for $u \in \hat{C}_0^\infty$, using again the Fubini-Tonelli theorem to justify the interchange in the order of integration,

$$Au(x) = \mathcal{F}^{-1}(a(\xi)\hat{u}(\xi))(x) = (2\pi)^{-d}\int e^{i<x,\xi>}\int e^{-i<y,\xi>}d\mu(y)\hat{u}(\xi)d\xi$$

$$= \int (2\pi)^{-d} \int e^{i<x-y,\xi>}\hat{u}(\xi)d\xi d\mu(y) = \int u(x-y)d\mu(y) = u*\mu(x).$$

This proves (2.5), and

$$\|Au\|_\infty \le V(\mu)\|u\|_\infty .$$

Hence $a \in M_\infty$, and the equality (2.4) follows as before.

This completes the proof of the theorem.

In particular, if $a \in M_\infty$, and if μ is as in the theorem, we may define a bounded linear operator A on L_∞ with norm $M_\infty(a)$ by $Au = u * \mu$ for $u \in L_\infty$. It is easily seen that if a is slowly increasing, then we have in the sense of distributions, $Au = \mathcal{F}^{-1}(a\hat{u})$, for $u \in L_\infty$, so that A coincides on L_∞ with the extension to S' of the operator in (1.2) on $\hat{C}_0^\infty$.

Our next two results describe inclusions and norm relations among different spaces of multipliers. The proofs will be based on the following well known lemma.

Lemma 2.1. (The Riesz-Thorin interpolation theorem.) Let $1 \le p_0, p_1, r_0, r_1 \le \infty$ and let T be a linear operator from $L_{p_0} \cap L_{p_1}$ into $L_{r_0} \cap L_{r_1}$ such that there exist constants N_0 and N_1 such that

$$\|Tf\|_{r_i} \le N_i \|f\|_{p_i}, \text{ for } f \in L_{p_i}, \; i = 0,1.$$

Let $0 < \theta < 1$ and let p and r be defined by

$$\frac{1}{p} = \frac{1-\theta}{p_0} + \frac{\theta}{p_1}, \quad \frac{1}{r} = \frac{1-\theta}{r_0} + \frac{\theta}{r_1} .$$

Then T may be extended to a bounded linear operator from L_p to L_r with

$$\|Tf\|_r \le N_0^{1-\theta} N_1^\theta \|f\|_p, \text{ for } f \in L_p .$$

Theorem 2.4. Let $1/p + 1/p' = 1$ with $1 \le p \le p' \le \infty$ and assume that $a \in M_p$. Then $a \in M_q$ for all q with $p \le q \le p'$ and

(2.7) $\qquad M_q(a) \le M_p(a).$

In particular, if $a \in M_p$ for some p with $1 \leq p \leq \infty$, then a is bounded and

$$\|a\|_\infty \leq M_p(a).$$

Proof. By Theorem 2.1 we have $a \in M_{p'}$ with $M_{p'}(a) = M_p(a)$, and hence the operator A in (1.2) is bounded in both L_p and $L_{p'}$. Writing $1/q = (1-\theta)/p + \theta/p'$ we therefore obtain by Lemma 2.1, for $u \in \hat{C}_0^\infty$,

$$\|Au\|_q \leq M_p(a)^{1-\theta} M_{p'}(a)^\theta \|u\|_q = M_p(a)\|u\|_q ,$$

which proves (2.7). The last statement is now an immediate consequence of Theorem 2.2.

Theorem 2.5. Assume that $a \in M_\infty$. Then $a \in M_p$ for $1 \leq p \leq \infty$ and

$$M_p(a) \leq M_2(a)^{2-2/p} M_\infty(a)^{2/p-1}, \quad \text{for } p \leq 2,$$

$$M_p(a) \leq M_2(a)^{2/p} M_\infty(a)^{1-2/p}, \quad \text{for } p \geq 2 .$$

Proof. The fact that $a \in M_p$ for $1 \leq p \leq \infty$ is contained in Theorem 2.4. Since $M_\infty(a) = M_1(a)$, the inequalities now follow by applying Lemma 2.1 to the operator A in (1.2).

We shall now prove that under certain conditions, limits of multipliers are multipliers.

Theorem 2.6. Let $a_n \in M_p$, $n = 1,2,\ldots$ be such that for some constant K,

(2.8) $$M_p(a_n) \leq K, \quad n = 1,2,\ldots .$$

Assume further that there exists a function $a \in C^\infty$ such that for every $v \in C_0^\infty$,

(2.9) $$\lim_{n \to \infty} \int a_n v \, d\xi = \int a v \, d\xi .$$

Then $a \in M_p$ and $M_p(a) \leq K$.

Proof. Setting $A_n u = \mathcal{F}^{-1}(a_n \hat{u})$ we obtain by (2.9) for each $u \in \hat{C}_0^\infty$ and $x \in R^d$,

$$\lim_{n\to\infty} A_n u(x) = (2\pi)^{-d} \lim_{n\to\infty} \int a_n(\xi) e^{i<x,\xi>} \hat{u}(\xi) d\xi = \mathcal{F}^{-1}(a\hat{u})(x) = Au(x).$$

Further, since by Theorems 2.2 and 2.4,

$$\|a_n\|_\infty \leq M_p(a_n) \leq K,$$

we have

$$|A_n u(x)| \leq (2\pi)^{-d} \|a_n\|_\infty \|\hat{u}\|_1 \leq (2\pi)^{-d} K \|\hat{u}\|_1,$$

and hence by dominated convergence, for $v \in C_0^\infty$,

$$\lim_{n\to\infty} \int A_n u \cdot v \, dx = \int Au \cdot v \, dx. \tag{2.10}$$

On the other hand, using Hölder's inequality, we have for p and p' conjugate indices,

$$\left| \int A_n u \cdot v dx \right| \leq \|A_n u\|_p \|v\|_{p'},$$

so that by (2.8) and (2.10),

$$\left| \int Au \cdot v dx \right| \leq K \|u\|_p \|v\|_{p'}.$$

The converse of Hölder's inequality then proves that

$$\|Au\|_p \leq K \|u\|_p,$$

which completes the proof of the theorem.

We next show that M_p is closed under multiplication.

Theorem 2.7. Let $a, b \in M_p$. Then $ab \in M_p$ and

$$M_p(ab) \leq M_p(a) M_p(b).$$

Proof. Let $u \in \hat{C}_0^\infty$. Then $Bu \equiv \mathcal{F}^{-1}(b\hat{u}) \in \hat{C}_0^\infty$ and hence with the notation (1.2),

$$\|A(Bu)\|_p \le M_p(a)\|Bu\|_p \le M_p(a)M_p(b)\|u\|_p .$$

Since $A(Bu) = \mathcal{F}^{-1}(ab\,\hat{u})$, this proves the theorem.

In the next two theorems we shall study the behavior of multipliers under affine transformations. It will be convenient to prove first the following lemma, in which we denote $(a \otimes b)(\xi,\eta) = a(\xi)b(\eta)$.

Lemma 2.2. Let $a \in M_p^{(l)}$, $b \in M_p^{(n)}$. Then $a \otimes b \in M_p^{(l+n)}$ and

$$M_p^{(l+n)}(a \otimes b) = M_p^{(l)}(a)M_p^{(n)}(b).$$

In particular, if $a \in M_p^{(l)}$, the natural extension $a \otimes 1$ of a to R^{l+n} is in $M_p^{(l+n)}$ and

$$M_p^{(l+n)}(a \otimes 1) = M_p^{(l)}(a).$$

Proof. Let (x,y) denote the variable in R^{l+n} and (ξ,η) its dual variable. Consider first the extension $\tilde{a}(\xi,\eta) \equiv a(\xi) \otimes 1$ of a to a function on R^{l+n}. We then have, with obvious notation, for $u \in \hat{C}_0^\infty(R^{l+n})$,

$$\tilde{A}u(x,y) = \mathcal{F}^{-1}_{\xi,\eta}(\tilde{a}\mathcal{F}_{x,y}u)(x,y) = \mathcal{F}^{-1}_{\xi}(a(\xi)\mathcal{F}^{-1}_{\eta}\mathcal{F}_y\mathcal{F}_x u)(x,y)$$

$$= \mathcal{F}^{-1}_{\xi}(a(\xi)\mathcal{F}_x u)(x,y).$$

Integration with respect to x gives

$$\int |\tilde{A}u(x,y)|^p dx \le M_p^{(l)}(a)^p \int |u(x,y)|^p dx ,$$

and after integration also with respect to y we conclude that $\tilde{a} \in M_p^{(l+n)}$ and

$$M_p^{(l+n)}(\tilde{a}) \le M_p^{(l)}(a).$$

Denoting similarly $\tilde{b}(\xi,\eta) = 1 \otimes b(\eta)$ we conclude by Theorem 2.7 that

$a \otimes b = \tilde{a}\tilde{b} \in M_p^{(l+n)}$ and

(2.11) $\qquad M_p^{(l+n)}(a \otimes b) \leq M_p^{(l)}(a) M_p^{(n)}(b).$

In order to prove equality, we take $u(x,y) = v(x)w(y)$ with $v \in \hat{C}_0^{\infty}(R^l)$, $w \in \hat{C}_0^{\infty}(R^n)$ and obtain

$$\mathcal{F}^{-1}_{\xi,\eta}(a \otimes b \mathcal{F}_{x,y} u)(x,y) = \mathcal{F}^{-1}_{\xi}(a \mathcal{F}_x v)(x) \cdot \mathcal{F}^{-1}_{\eta}(b \mathcal{F}_y w)(y),$$

form which at once, by letting v and w vary,

$$M_p^{(l+n)}(a \otimes b) \geq M_p^{(l)}(a) M_p^{(n)}(b).$$

Together with (2.11) this completes the proof.

For T an affine mapping of R^d into R^m and a function a on R^m, we shall denote by a_T the function on R^d defined by $a_T(\xi) = a(T\xi)$.

Theorem 2.8. Let $m \leq d$ and let T be an affine surjection of R^d onto R^m. Then if $a \in M_p^{(m)}$ we have that $a_T \in M_p^{(d)}$ and

$$M_p^{(d)}(a_T) = M_p^{(m)}(a).$$

Proof. By Theorem 2.1 it suffices to consider the case $1 \leq p \leq 2$. Since

$$\mathcal{F}^{-1}(a(\cdot - \omega)\hat{u})(x) = e^{i\langle x,\omega\rangle} \mathcal{F}^{-1}(a \mathcal{F}(e^{-i\langle x,\omega\rangle} u))(x),$$

it follows that a translation is an isometry of M_p, and it therefore suffices to assume that T is linear.

Assume first that $m = d$ and thus that T is non-singular. Letting $'$ denote transposition we have for $u \in \hat{C}_0^{\infty}(R^d)$,

$$A_T u(x) \equiv (2\pi)^{-d} \int e^{i\langle x,\xi\rangle} a(T\xi)\hat{u}(\xi) d\xi = |\det(T^{-1})| \mathcal{F}^{-1}(a\hat{u}(T^{-1}\cdot))((T^{-1})'x).$$

Hence we obtain via another transformation of variables,

$$(2.12) \qquad \|A_T u\|_p^p \leq |\det(T^{-1})|^p |\det(T')| M_p(a) \|\mathcal{F}^{-1}(\hat{u}(T^{-1}\cdot))\|_p^p .$$

By a similar calculation we find that

$$\mathcal{F}^{-1}(\hat{u}(T^{-1}\cdot))(x) = |\det(T)| u(T'x) ,$$

and hence

$$\|\mathcal{F}^{-1}(\hat{u}(T^{-1}\cdot))\|_p^p = |\det(T)|^p |\det((T^{-1})')| \|u\|_p^p .$$

Inserting this into (2.12) we find that

$$\|A_T u\|_p^p \leq M_p(a)^p \|u\|_p^p ,$$

which proves that $a_T \in M_p$ and

$$M_p(a_T) \leq M_p(a) .$$

Since we have assumed that T is invertible, the opposite inequality is proved in the same way. This proves the theorem in the case $m = d$.

Consider now an arbitrary linear map T of R^d onto R^m, and decompose it as $T = T_1 P$, where T_1 is a non-singular map on R^m and P is the natural projection of R^d onto the subspace determined by its first m coordinates. We then have $a_T = (a_{T_1})_P$ and since a_T is simply the extension $a_{T_1} \otimes 1$ of a_{T_1} from R^m to R^d we have by Lemma 2.2 and the result already proved,

$$M_p^{(d)}(a_T) = M_p^{(m)}(a_{T_1}) = M_p^{(m)}(a) .$$

This completes the proof of the theorem.

We finally prove that restrictions of multipliers to hyperplanes are again multipliers "on the hyperplane" in the following sense.

Theorem 2.9. Let $m > d$ and let T be an affine injection of R^d into R^m. Then if $a \in M_p^{(m)}$ we have that $a_T \in M_p^{(d)}$ and

$$M_p^{(d)}(a_T) \leq M_p^{(m)}(a).$$

Proof. By Theorem 2.8 it is sufficient to consider the case that T is the natural injection of R^d onto the subspace of R^m defined by its first d coordinates so that if $\eta = (\xi,\zeta) \in R^m$ with $\xi \in R^d$, we have $T\xi = (\xi,0)$ and $a_T(\xi) = a(\xi,0)$. Setting $a_n(\eta) = a(\xi,n^{-1}\zeta)$, $n = 1,2,\ldots,$ we have by Theorem 2.8,

$$M_p^{(m)}(a_n) = M_p^{(m)}(a) = \text{constant}.$$

Since a_n tends to $a(\xi, 0)$ uniformly on compact sets in R^m as n tends to infinity, we may apply Theorem 2.6 to conclude that $a(\xi,0) \in M_p^{(m)}$ and

$$M_p^{(m)}(a(\xi,0)) \leq M_p^{(m)}(a).$$

Since $a(\xi,0)$ is just the extension $a_T(\xi) \otimes 1$ of a_T from R^d to R^m, this completes the proof by Lemma 2.2.

1.3. The Carlson-Beurling inequality.

The elementary bound for the M_∞ norm given in this section will be one of our main technical tools in later applications.

For convenience we introduce the norm in the homogeneous Sobolev space $\dot{W}_2^m$ and its inhomogeneous counterpart W_2^m, viz. $(|\alpha| = \alpha_1 + \ldots + \alpha_d)$

$$\|v\|_{\dot{W}_2^m} = \sum_{|\alpha|=m} \|D^\alpha v\|_2,$$

$$\|v\|_{W_2^m} = \sum_{|\alpha|\leq m} \|D^\alpha v\|_2 .$$

We recall Sobolev's lemma.

Lemma 3.1. Let ν be an integer with $\nu > d/2$. Then there exists a constant C such that

$$\|v\|_\infty \leq C\|v\|_{W_2^\nu} .$$

In particular, for $\nu > d/2$ the norm in W_2^ν dominates the $M_2^{(d)}$ norm. We shall see that in fact it even dominates the norm in $M_\infty^{(d)}$.

Theorem 3.1. (The Carlson-Beurling inequality.) Let ν be an integer with $\nu > d/2$. Then there exists a constant C such that for any $a \in C^\infty$ with $\|a\|_{W_2^\nu} < \infty$, we have $a \in M_\infty^{(d)}$ and

$$(3.1) \qquad M_\infty^{(d)}(a) \leq C\|a\|_2^{1-d/(2\nu)}\|a\|_{\dot{W}_2^\nu}^{d/(2\nu)} .$$

Proof. Assume first that $a \in C_0^\infty$, $a \not\equiv 0$. Then $\check{a} \in L_1$ and by Theorem 2.3 we have $M_\infty(a) = \|\check{a}\|_1$. Let $\omega > 0$. By Schwarz' inequality and Parseval's formula,

$$\|\check{a}\|_1 = \int_{|x|\geq\omega} |x|^{-\nu}|x|^\nu|\check{a}(x)|\,dx + \int_{|x|\leq\omega} |\check{a}(x)|\,dx$$

$$\leq C\{(\int_{|x|\geq\omega} |x|^{-2\nu}dx)^{1/2}(\int |x|^{2\nu}|\check{a}(x)|^2dx)^{1/2} + \omega^{d/2}(\int_{|x|\leq\omega} |\check{a}(x)|^2dx)^{1/2}\}$$

$$\leq C\{\omega^{-(2\nu-d)/2}\|a\|_{\dot{W}_2^\nu} + \omega^{d/2}\|a\|_2\} .$$

The desired inequality follows by setting

$$\omega = (\|a\|_2/\|a\|_{\dot{W}_2^\nu})^{-1/\nu} .$$

Assume now that a does not have compact support. Then there is a sequence $\{a_n\}_1^\infty \subset C_0^\infty$ such that

$$\lim_{n\to\infty} \|a_n - a\|_{W_2^\nu} = 0.$$

Applying (3.1) to a_n we have for n large enough,

$$M_\infty^{(d)}(a_n) \le C\|a_n\|_2^{1-d/(2\nu)}\|a_n\|_{\dot{W}_2^\nu}^{d/(2\nu)} \le 2C\|a\|_2^{1-d/(2\nu)}\|a\|_{\dot{W}_2^\nu}^{d/(2\nu)} .$$

The result therefore follows by Theorem 2.6 since, by Sobolev's lemma, the a_n converge uniformly to a as n tends to infinity.

1.4. Periodic multipliers.

In this section we shall prove that the M_p norm of a C^∞ periodic function can be estimated by the M_p norm of an associated function with compact support. Combined with the Carlson-Beurling inequality this will provide a convenient tool in our applications for deriving precise estimates for periodic multipliers depending on parameters.

By Theorem 2.8 we may normalize the period to be 2π in each variable. We notice that any C^∞ function which is 2π-periodic can be expanded in an absolutely convergent Fourier series,

$$a(\xi) = \sum_{\alpha \in Z^d} a_\alpha e^{-i\langle\alpha,\xi\rangle}, \quad \sum_\alpha |a_\alpha| < \infty ,$$

where with $Q = \{\xi: |\xi_j| \le \pi, j = 1,\dots,d\}$, the Fourier coefficients are given by

$$(4.1) \qquad a_\alpha = (2\pi)^{-d} \int_Q e^{i\langle\alpha,\xi\rangle} a(\xi) d\xi .$$

Denoting by $\mu_a \in B$ the measure with point mass a_α at $\alpha \in Z^d$, we may write

$$a(\xi) = \int e^{-i\langle x,\xi\rangle} d\mu_a(x), \quad V(\mu_a) = \sum_\alpha |a_\alpha| ,$$

and Theorem 2.3 shows that $a \in M_\infty$, and hence by Theorem 2.4 that $a \in M_p$ for $1 \le p \le \infty$. The translation invariant operator A associated with a by (1.2) is here

$$(4.2) \qquad Au(x) = u * \mu_a(x) = \sum_{\alpha \in Z^d} a_\alpha u(x-\alpha), \text{ for } u \in \hat{C}_0^\infty .$$

The following is now the main result of this section.

Theorem 4.1. Let $C_0^\infty \ni \eta : R^d \to [0,1]$ be identically equal to unity in a neighborhood of Q and have its support in the interior of $(5/4)Q$. Then there exists a constant C such that for any 2π-periodic function a in C^∞,

$$M_p(a) \leq CM_p(\eta a).$$

The proof, which is postponed till the end of this section, will be based on the following lemma where for any $f \in C_0^\infty$ we denote by $\tilde{f}$ the periodic function defined by

$$\tilde{f}(\xi) = \sum_{\alpha \in Z^d} f(\xi + 2\pi\alpha).$$

Lemma 4.1. Let $b \in C_0^\infty(R^d)$. There exists a constant C depending only on the diameter of the support of b such that

$$M_p(\tilde{b}) \leq CM_p(b).$$

Proof. By Theorem 2.1 it is enough to consider the case $p < \infty$ (in fact, $1 \leq p \leq 2$). We shall assume first that b has its support in $\text{int}(Q)$, the interior of Q. In this case $\tilde{b}$ is the periodic extension of b.

For multi-sequences $c = \{c_\alpha\}_{\alpha \in Z^d}$, let

$$\|c\|_{l_p} \equiv \|\{c_\cdot\}\|_{l_p} = \Big(\sum_{\alpha \in Z^d} |c_\alpha|^p \Big)^{1/p} .$$

With b_α the Fourier coefficients of the periodic function $\tilde{b}$, we then define the discrete multiplier norm

$$m_p(b) = \sup\{ \|\{\sum_\alpha b_\alpha c_{\cdot - \alpha}\}\|_{l_p} : \|c\|_{l_p} \leq 1, \; c \text{ finite}\} .$$

Letting $\tilde{B}$ be the operator associated by (1.2) with $\tilde{b}$, we have by (4.1) and (4.2), for $u \in \hat{C}_0^\infty$,

$$\tilde{B}u = \sum_\alpha b_\alpha u(\cdot-\alpha) ,$$

and hence, for every $x \in R^d$,

$$\|\{\tilde{B}u(x-\cdot)\}\|_{l_p} \leq m_p(b)\|\{u(x-\cdot)\}\|_{l_p} .$$

It follows, with $Q_1 = \{x: 0 \leq x_j \leq 1,\ j = 1,\dots,d\}$ that

$$\|\tilde{B}u\|_p^p = \sum_\beta \int_{Q_1} |\tilde{B}u(x-\beta)|^p dx = \int_{Q_1} \|\{\tilde{B}u(x-\cdot)\}\|_{l_p}^p dx$$

$$\leq m_p(b)^p \int_{Q_1} \|\{u(x-\cdot)\}\|_{l_p}^p dx = m_p(b)^p\|u\|_p^p ,$$

so that

(4.3) $$M_p(\tilde{b}) \leq m_p(b) .$$

For the purpose of estimating $m_p(b)$ by $CM_p(b)$ we shall need to associate multi-sequences with functions on R^d and conversely. For a function f on R^d we set $Pf = \{P_\alpha f\}$ with

(4.4) $$P_\alpha f = \int_{Q_1+\alpha} f(x)dx \quad \text{for } \alpha \in Z^d,$$

and for a sequence $c = \{c_\alpha\}$ we define the extension Ec to R^d by

(4.5) $$Ec(x) = \sum_\alpha c_\alpha \chi(x-\alpha) \quad \text{for } x \in R^d,$$

where χ is the characteristic function of Q_1. We then have

(4.6) $$\|Pf\|_{l_p} \leq \|f\|_p$$

and

(4.7) $$\|Ec\|_p = \|c\|_{l_p} .$$

Introducing the Fourier series corresponding to $c = \{c_\alpha\}$,

$$\hat{c}(\xi) = \sum_\alpha c_\alpha e^{-i<\xi,\alpha>} ,$$

we find at once by (4.5),

$$\mathcal{F}(Ec)(\xi) = \hat{c}(\xi)\hat{\chi}(\xi) ,$$

where

$$\hat{\chi}(\xi) = \int_{Q_1} e^{-i<x,\xi>}dx = \prod_{j=1}^{d} \left(\frac{1-e^{-i\xi_j}}{i\xi_j}\right) .$$

Let now $\phi \in C_0^\infty(R^d)$ be such that

$$\phi(\xi)|\hat{\chi}(\xi)|^2 = 1 \quad \text{for} \quad \xi \in Q .$$

Since $\text{supp}(b) \subset \text{int}(Q)$ we have

$$b_\alpha \equiv (2\pi)^{-d} \int_Q e^{i<\alpha,\xi>}\tilde{b}(\xi)d\xi = \mathcal{F}^{-1} b\,(\alpha) ,$$

and we may then write

$$b_\alpha = \mathcal{F}^{-1}(b\,\phi\,\hat{\chi}\,\bar{\hat{\chi}})(\alpha) = \int_{Q_1} \mathcal{F}^{-1}(b\,\phi\,\hat{\chi})(\alpha+x)dx .$$

Hence for $c = \{c_\beta\}$ finite,

$$\sum_\beta c_\beta b_{\alpha-\beta} = \int_{Q_1} \sum_\beta c_\beta \mathcal{F}^{-1}(b\phi\,\hat{\chi})(\alpha-\beta+x)dx = \int_{Q_1+\alpha} \mathcal{F}^{-1}(b\phi\,\hat{c}\,\hat{\chi})(x)dx .$$

Therefore, introducing the operator in L_1 defined by $B_\phi u = \mathcal{F}^{-1}(b\phi\,\hat{u})$, and using the notation (4.4), (4.5) we can write, since $Ec \in L_1$,

$$\sum_\beta c_\beta b_{\alpha-\beta} = P_\alpha(B_\phi(Ec)).$$

By (4.6) and (4.7) this implies

$$\|\{\sum_\beta c_\beta b_{\cdot-\beta}\}\|_{l_p} = \|\mathbb{P}(B_\phi(Ec))\|_{l_p} \leq \|B_\phi(Ec)\|_p \leq M_p(b\phi)\|Ec\|_p \leq M_p(\phi)M_p(b)\|c\|_{l_p} ,$$

so that

$$m_p(b) \leq M_p(\phi)M_p(b).$$

In view of (4.3) this proves the lemma in the case $\operatorname{supp}(b) \subset \operatorname{int}(Q)$.

Consider now the general case where $\operatorname{supp}(b)$ is not necessarily contained in $\operatorname{int}(Q)$ and let $\eta_1,\dots,\eta_l \in C_0^\infty$ be such that $\eta_1 + \dots + \eta_l = 1$ on $\operatorname{supp}(b)$ and such that there exists $\xi_j \in R^d$, $j = 1,\dots,l$, such that

$$\operatorname{supp}(\eta_j(\cdot-\xi_j)) \subset \operatorname{int}(Q).$$

Then since $\tilde{b} = \sum_j \widetilde{\eta_j b}$ we obtain, using also Theorems 2.8 and 2.7,

$$M_p(\tilde{b}) \leq \sum_j M_p(\widetilde{\eta_j b}) = \sum_j M_p((\widetilde{\eta_j b})(\cdot-\xi_j)) \leq C \sum_j M_p(\eta_j b) \leq CM_p(b) .$$

Since the η_j can be chosen as translates of a fixed function and the number l of η_j:s then only has to depend on the diameter of $\operatorname{supp}(b)$, this completes the proof of the lemma.

We now prove Theorem 4.1.

<u>Proof</u>. The periodicity of a implies that $(\widetilde{\eta a}) = \tilde{\eta}a$ and hence Lemma 4.1 gives

$$(4.8) \qquad M_p(\tilde{\eta}a) = M_p((\widetilde{\eta a})) \leq CM_p(\eta a).$$

Since $\tilde{\eta} \in C^\infty$ is strictly positive we have that $1/\tilde{\eta} \in C^\infty$. Periodicity hence yields $1/\tilde{\eta} \in M_p$ and it follows that

$$M_p(a) = M_p(\frac{1}{\tilde{\eta}} \cdot \tilde{\eta}a) \leq CM_p(\tilde{\eta}a).$$

If we combine this with (4.8), the theorem is proved.

1.5. Van der Corput's lemma.

In this section we shall prove a lemma by van der Corput and apply it to derive lower bounds for the norms of some special multipliers.

Lemma 5.1. (van der Corput's lemma.) Let $\Phi \in C^2(R^1)$ be real with $|\Phi''(\xi)| \geq \delta > 0$ on an interval $[a,b]$. Then

$$|\int_a^b \exp(i\Phi)d\xi| \leq 8\delta^{-1/2} .$$

Proof. We may assume, without loss of generality, that $\Phi'' \geq \delta > 0$ on $[a,b]$ so that Φ' is strictly increasing. Assume first that $\Phi' \geq 0$ on $[a,b]$. Let $a < c < b$ with c to be chosen later, and consider

$$I_c = \int_c^b \exp(i\Phi)d\xi = -i \int_c^b \frac{1}{\Phi'} \frac{d}{d\xi} \exp(i\Phi)d\xi .$$

Recall the second mean value theorem in the following form: Let $f,g \in C^1[c,b]$ with f non-negative decreasing and g real, $|g| \leq M$. Then

$$|\int_c^b fg'd\xi| \leq 2Mf(c) .$$

For a proof, we integrate by parts to obtain

$$\pm \int_c^b fg'd\xi = \pm f(b)(g(b) - g(c)) \mp \int_c^b (g(\xi) - g(c))f'(\xi)d\xi$$

$$\leq 2Mf(b) + 2M(f(c) - f(b)) = 2Mf(c).$$

Applying this with $f = 1/\Phi'$ and $g = \cos \Phi$ and $\sin \Phi$ we obtain, since $\Phi'(\xi) \geq (c-a)\delta$ on $[c,b]$, that

$$|I_c| \leq |\mathrm{Re}\, I_c| + |\mathrm{Im}\, I_c| \leq \frac{4}{(c-a)\delta} .$$

Hence

$$\left|\int_a^b \exp(i\Phi)d\xi\right| \le |I_c| + \left|\int_a^c \exp(i\Phi)d\xi\right| \le \frac{4}{(c-a)\delta} + c-a .$$

If $2\delta^{-1/2} < b-a$ we choose $c = a + 2\delta^{-1/2}$ and obtain

$$\left|\int_a^b \exp(i\Phi)d\xi\right| \le 4\delta^{-1/2} .$$

Since this latter inequality is trivial if $2\delta^{-1/2} \ge b-a$, this proves the lemma if $\Phi' \ge 0$ on $[a,b]$.

If instead $\Phi' \le 0$, the result follows by applying the above to the function $\Phi(-\xi)$. In the general case, $[a,b]$ is the union of two intervals where Φ' has constant sign and the lemma now follows by adding the corresponding two estimates.

As a consequence we have the following result:

Lemma 5.2. Let $g \in C_0^\infty(R^1)$ and let $\Phi \in C^2(R^1)$ be real with $|\Phi''| \ge \delta > 0$ in a neighborhood of $\operatorname{supp}(g)$. Then there exists a constant C depending only on $\operatorname{supp}(g)$ such that for $t > 0$,

$$\|\mathcal{F}^{-1}(g \exp(it\Phi))\|_\infty \le C\delta^{-1/2}t^{-1/2}\|g'\|_1 .$$

Proof. Since $\operatorname{supp}(g)$ is compact it may be covered by a finite number of intervals on which $|\Phi''| \ge \delta > 0$, and hence it suffices to prove the lemma under the assumption that $|\Phi''| \ge \delta > 0$ on an interval containing $\operatorname{supp}(g)$. If $c \in \operatorname{supp}(g)$ we have

$$\mathcal{F}^{-1}(g \exp(it\Phi))(x) = (2\pi)^{-1}\int g(\xi)\exp(ix\xi + it\Phi(\xi))d\xi$$

$$= -(2\pi)^{-1}\int g'(\xi)\left(\int_c^\xi \exp(ixy + it\Phi(y))dy\right)d\xi .$$

Lemma 5.1 applied to the inner integral now proves the desired result.

We shall now apply Lemma 5.2 to obtain lower bounds for the norms of certain multipliers.

Corollary 5.1. Let $g \in C_0^\infty(R^1)$ and let $\Phi \in C^\infty(R^1)$ be real and such that $\Phi'' \not\equiv 0$ at some point where $g \not\equiv 0$. Then there exists a constant $c > 0$ such that for $t > 0$,

$$M_p(g \exp(it\Phi)) \geq ct^{\left|\frac{1}{2}-\frac{1}{p}\right|}.$$

Proof. We may assume that $p \leq 2$. Let $\chi \in C_0^\infty$ be real with $\operatorname{supp}(\chi) \subset \{x : g(x) \not\equiv 0\}$ and $\Phi'' \not\equiv 0$ on $\operatorname{supp}(\chi)$. Then $\chi/g \in C_0^\infty \subseteq M_p$ and hence

$$M_p(\chi \exp(it\Phi)) \leq M_p(\chi/g) M_p(g \exp(it\Phi)) = CM_p(g \exp(it\Phi)) . \tag{5.1}$$

Let $w \in C_0^\infty$ with $w = 1$ on $\operatorname{supp}(\chi)$. By Hölder's inequality and Parseval's formula we have with $1/p + 1/p' = 1$,

$$\begin{aligned} 0 < \|\chi\|_2^2 &= \int \chi \exp(it\Phi) w \chi \exp(-it\Phi) d\xi \\ &= 2\pi \int \mathcal{F}^{-1}(\chi \exp(it\Phi)w)\overline{\mathcal{F}^{-1}(\chi \exp(it\Phi))}dx \\ &\leq 2\pi \|\mathcal{F}^{-1}(\chi \exp(it\Phi)w)\|_p \|\mathcal{F}^{-1}(\chi \exp(it\Phi))\|_{p'} \\ &\leq 2\pi M_p(\chi \exp(it\Phi)) \|\check{w}\|_p \|\mathcal{F}^{-1}(\chi \exp(it\Phi))\|_{p'} . \end{aligned} \tag{5.2}$$

Again using Hölder's inequality and Parseval's formula we obtain

$$\begin{aligned} \|\mathcal{F}^{-1}(\chi \exp(it\Phi))\|_{p'} &\leq \|\mathcal{F}^{-1}(\chi \exp(it\Phi))\|_2^{2/p'} \|\mathcal{F}^{-1}(\chi \exp(it\Phi))\|_\infty^{1-2/p'} \\ &= (2\pi)^{-1/p'} \|\chi\|_2^{2/p'} \|\mathcal{F}^{-1}(\chi \exp(it\Phi))\|_\infty^{2/p-1} . \end{aligned}$$

By (5.1), (5.2) and Lemma 5.2, it hence follows that

$$0 < \|\chi\|_2^2 \leq CM_p(g \exp(it\Phi)) \|\mathcal{F}^{-1}(\chi \exp(it\Phi))\|_\infty^{2/p-1} \leq Ct^{\left(\frac{1}{2}-\frac{1}{p}\right)} M_p(g \exp(it\Phi)) ,$$

which proves the corollary.

By restriction to a suitable hyperplane, Corollary 5.1 implies the following analogous estimate in several dimensions. This estimate can in fact be improved in certain cases, see e.g. Lemma 6.1.1.

Corollary 5.2. Let $g \in C_0^\infty(R^d)$ and let $\Phi \in C^\infty(R^d)$ be real and such that not all the second derivatives of Φ vanish identically on $\mathrm{supp}(g)$. Then there exists a constant $c > 0$ such that for $t > 0$,

$$M_p(g\exp(it\Phi)) \geq ct^{|\frac{1}{2}-\frac{1}{p}|} .$$

Proof. Since an orthogonal change of variables does not change the multiplier norm (Theorem 2.8) we may assume that $\partial^2\Phi/\partial\xi_1^2$ does not vanish identically on $\mathrm{supp}(g)$. Let ξ^0 be a point where both g and $\partial^2\Phi/\partial\xi_1^2$ are non-zero, and set

$$\tilde{\Phi}(\xi_1) = \Phi(\xi_1,\xi_2^0,\ldots,\xi_d^0), \quad \tilde{g}(\xi_1) = g(\xi_1,\xi_2^0,\ldots,\xi_d^0) .$$

By Corollary 5.1 we have

$$(5.3) \qquad M_p^{(1)}(\tilde{g}\exp(it\tilde{\Phi})) \geq ct^{|\frac{1}{2}-\frac{1}{p}|} ,$$

and by Theorem 2.9 with $T:R^1 \to R^d$ the injection $\xi_1 \sim (\xi_1,\xi_2^0,\ldots,\xi_d^0)$,

$$M_p^{(1)}(\tilde{g}\exp(it\tilde{\Phi})) \leq M_p^{(d)}(g\exp(it\Phi)).$$

Combined with (5.3) this proves the corollary.

Finally we use the last result to prove that certain bounded smooth functions are multipliers on L_p only for $p = 2$.

Corollary 5.3. Let P be a not identically vanishing homogeneous real polynomial on R^d of degree $\nu > 1$. Then $\exp(iP)$ does not belong to M_p for $p \neq 2$.

Proof. Assume, in order to reach a contradiction, that $\exp(iP) \in M_p$ for some $p \neq 2$. By homogeneity, $t^\nu P(\xi) = P(t\xi)$ and hence by Theorem 2.8,

$$M_p(\exp(iP)) = M_p(\exp(it^\nu P)) .$$

Since P is not linear we may choose $g \in C_0^\infty$ such that not all the second order derivatives of P vanish identically on the support of g. Then for $t > 0$,

$$M_p(g\exp(it^\nu P)) \leq M_p(g)M_p(\exp(it^\nu P)) = \text{constant}.$$

On the other hand, by Corollary 5.2 the left hand side above tends to infinity with t. This contradiction establishes the corollary.

References.

For basic material on measure theory, in particular L_p spaces, the Fubini-Tonelli theorem and the Riesz representation theorem, see [7]. A proof of the Riesz-Thorin interpolation theorem is given in [8, Chapter 12]. For the distribution theory used in this chapter, cf. e.g. [4], where also a proof of Sobolev's lemma can be found ([4, Theorem 2.2.7]).

The material in Section 2 is taken mainly from [3]. The one-dimensional Carlson-Beurling inequality appears in [1]. Our proof of Lemma 4.1 is based on the presentation in [5]; in the case $p = 1$ or ∞, a simple proof is given in [6, Theorem 2.7.6]. van der Corput's lemma is contained in [2].

1. A. Beurling, Sur les intégrales de Fourier absolument convergentes et leur applications à une transformation functionelle, IX Congrès des Math. Scandinave, Helsingfors (1935), 345-366.

2. J.G. van der Corput, Zur Methode der Stationären Phase I, Compositio Math. 1 (1934), 15-38.

3. L. Hörmander, Estimates for translation invariant operators in L_p spaces, Acta Math. 104 (1960), 93-140.

4. L. Hörmander, Linear Partial Differential Operators, Springer, Berlin 1969.

5. M. Jodeit, Restrictions and extensions of Fourier multipliers, Studia Math. 34 (1970), 215-226.

6. W. Rudin, Fourier Analysis on Groups, Interscience, New York 1962.

7. W. Rudin, Real and Complex Analysis, McGraw - Hill, New York 1966.

8. A. Zygmund, Trigonometric Series II, Cambridge University Press, Cambridge 1959.

CHAPTER 2. BESOV SPACES.

In this chapter we introduce the Besov spaces $B_p^{s,q}$ $(s > 0,\ 1 \leq p,q \leq \infty)$ and derive a number of their properties, to the extent needed in subsequent applications. In Section 1 we define $B_p^{s,q}$ in terms of the growth of the Fourier transform of its elements. In Section 2 we then investigate the dependence of $B_p^{s,q}$ upon its parameters and compare $B_p^{m,q}$, for m integer, to the Sobolev space W_p^m. We also derive a sharp form of Sobolev's inequality. In Section 3 we show that $B_p^{s,q}$ can also be defined in terms of modulii of continuity of derivatives in L_p; this was how these spaces were originally introduced by Besov. In Section 4 we describe, in terms of their relation to the Besov spaces, two special examples of one-parameter families of functions which will be used later to elucidate our applications. In Section 5 we show that in a certain sense the space $B_p^{s,\infty}$ interpolates a pair $B_p^{s_l,q_l}$, $l = 0,1$, of such spaces if $s_0 < s < s_1$. Section 6 finally contains two technical lemmas involving Besov spaces and Fourier multipliers.

2.1. Definition.

In this section we shall define the Besov spaces in terms of growth properties of the Fourier transforms of their elements. For this purpose we first need the following lemma.

Lemma 1.1. There exists a non-negative function $\phi \in C_0^\infty(R^d)$ with $\operatorname{supp}(\phi) \subset \{\xi : \frac{1}{2} < |\xi| < 2\}$ such that

$$\sum_{j=-\infty}^{\infty} \phi(2^{-j}\xi) = 1, \text{ for } \xi \neq 0 .$$

Proof. Letting $\tilde{\phi} \in C_0^\infty(R^d)$ be positive for $1/\sqrt{2} \le |\xi| \le \sqrt{2}$ and have $\mathrm{supp}(\tilde{\phi}) \subset \{\xi : \frac{1}{2} < |\xi| < 2\}$, the result follows at once by setting

$$\phi(\xi) = \tilde{\phi}(\xi)/(\sum_{j=-\infty}^{\infty} \tilde{\phi}(2^{-j}\xi)),$$

Throughout the rest of these notes we shall use the notation

$$\phi_j(\xi) = \phi(2^{-j}\xi), \text{ for } j \in Z,$$

where ϕ is the function in Lemma 1.1. Notice that since $\phi \in C_0^\infty \subset M_p$, $\mathcal{F}^{-1}(\phi_j \hat{v}) = (\mathcal{F}^{-1}\phi_j) * v \in L_p$ for $v \in L_p$.

For sequences $c = \{c_j\}_{j \in \tilde{Z}}$ where $\tilde{Z} \subset Z$, let

$$\|c\|_{l_q(\tilde{Z})} = \begin{cases} (\sum_{j \in \tilde{Z}} |c_j|^q)^{1/q}, & \text{for } 1 \le q < \infty, \\ \sup_{j \in \tilde{Z}} |c_j|, & \text{for } q = \infty. \end{cases}$$

We say that $c \in l_q(\tilde{Z})$ if the corresponding norm is finite. For $\tilde{Z} = Z$ we shall often write l_q instead of $l_q(Z)$.

Let now $s > 0$, $1 \le p \le \infty$, $1 \le q \le \infty$. We say that v belongs to the Besov space $B_p^{s,q}$ if $v \in L_p$ and the sequence $\{2^{sj}\|\mathcal{F}^{-1}(\phi_j \hat{v})\|_p\}_{j \in Z}$ is in l_q. Setting

(1.1) $$\|v\|_{\dot{B}_p^{s,q}} = \|\{2^{sj}\|\mathcal{F}^{-1}(\phi_j \hat{v})\|_p\}\|_{l_q},$$

$B_p^{s,q}$ is then a normed linear space with respect to the norm

(1.2) $$\|v\|_p + \|v\|_{\dot{B}_p^{s,q}}.$$

In fact, $B_p^{s,q}$ is a Banach space. For let $\{v^k\}_1^\infty$ be a Cauchy sequence in $B_p^{s,q}$. Then, in particular, it is a Cauchy sequence in L_p and hence converges to some $v \in L_p$. Then, for each $j \in Z$, $v_j^k = 2^{sj}\|\mathcal{F}^{-1}(\phi_j \hat{v}^k)\|_p$ converges to

$v_j = 2^{sj}\|\mathcal{F}^{-1}(\phi_j\hat{v})\|_p$ and hence since $\|v^k\|_{\dot{B}_p^{s,q}} = \|\{v_j^k\}\|_{l_q}$ remains bounded, we can conclude that $\{v_j\} \in l_q$ or $v \in \dot{B}_p^{s,q}$.

In our applications it will be convenient to work with an alternative norm to that in (1.2). For this purpose, let, (this notation will be used throughout these notes)

$$\psi_j = \phi_j \quad \text{for } j > 0, \quad \psi_0 = 1 - \sum_{j=1}^{\infty} \phi_j .$$

We then have $\mathrm{supp}(\psi_0) \subset \{\xi : |\xi| < 2\}$ and $\sum_{j=0}^{\infty} \psi_j = 1$ for all $\xi \in R^d$. Denoting the set of non-negative integers by Z^{0+}, we set for $v \in L_p$,

$$(1.3) \qquad \|v\|_{B_p^{s,q}} = \|\{2^{sj}\|\mathcal{F}^{-1}(\psi_j\hat{v})\|_p\}\|_{l_q(Z^{0+})} .$$

This definition will be used also for $s = 0$.

We shall now prove that the norms defined in (1.2) and (1.3) are equivalent.

<u>Theorem 1.1</u>. Let $s > 0$, $1 \leq p \leq \infty$, $1 \leq q \leq \infty$. Then there exist positive constants c and C such that for $v \in L_p$,

$$(1.4) \qquad c\|v\|_{B_p^{s,q}} \leq \|v\|_p + \|v\|_{\dot{B}_p^{s,q}} \leq C\|v\|_{B_p^{s,q}} .$$

For $s = 0$, $1 \leq p \leq \infty$, there exist positive constants c and C such that for $v \in L_p$,

$$(1.5) \qquad c\|v\|_{B_p^{0,\infty}} \leq \|v\|_p \leq C\|v\|_{B_p^{0,1}} .$$

<u>Proof</u>. Letting Z^+ denote the positive integers, we have

$$\|v\|_{B_p^{s,q}} \leq \|\mathcal{F}^{-1}(\psi_0\hat{v})\|_p + \|\{2^{sj}\|\mathcal{F}^{-1}(\phi_j\hat{v})\|_p\}\|_{l_q(Z^+)} \leq M_p(\psi_0)\|v\|_p + \|v\|_{\dot{B}_p^{s,q}} ,$$

which proves the lower estimate in (1.4). In order to prove the upper estimate in

(1.4), let $1/q + 1/q' = 1$. Then since $\Sigma_j \psi_j = 1$,

$$(1.6) \qquad \|v\|_p \le \sum_0^\infty \|\mathcal{F}^{-1}(\psi_j \hat{v})\|_p \le \|\{2^{-sj}\}\|_{l_{q'}(Z^{0+})} \|v\|_{B_p^{s,q}} = C\|v\|_{B_p^{s,q}} .$$

Further, with $Z^{0-} = Z \setminus Z^+$ and using Theorem 1.2.8 and in the last step, (1.6),

$$\|v\|_{\dot{B}_p^{s,q}} \le \|\{2^{sj}\|\mathcal{F}^{-1}(\phi_j \hat{v})\|_p\}\|_{l_q(Z^{0-})} + \|v\|_{B_p^{s,q}}$$

$$\le M_p(\phi)\|v\|_p\|\{2^{sj}\}\|_{l_q(Z^{0-})} + \|v\|_{B_p^{s,q}} \le C\|v\|_{B_p^{s,q}} .$$

This completes the proof of (1.4).

For (1.5), note that

$$\|v\|_{B_p^{0,\infty}} = \sup_{j \ge 0} \|\mathcal{F}^{-1}(\psi_j \hat{v})\|_p \le \sup_{j \ge 0} M_p(\psi_j)\|v\|_p = C\|v\|_p ,$$

which proves the lower estimate. The upper estimate is contained in (1.6) with $s = 0$, $q = 1$, and $q' = \infty$.

This completes the proof of the theorem.

Notice that the second inequality in (1.5) enables us to define the Banach space $B_p^{0,1}$ as a subset of L_p.

2.2. Embedding results.

In this section we shall compare Besov spaces with different indices, and also relate the Besov spaces to Sobolev spaces. If B_1 and B_2 are two Banach spaces we shall write $B_1 \subset B_2$ to mean that B_1 is continuously embedded in B_2 so that for the corresponding norms,

$$\|v\|_{B_2} \le C\|v\|_{B_1} , \text{ for } v \in B_1 .$$

We start with a result on Besov spaces $B_p^{s,q}$ for fixed p.

Theorem 2.1. Let $1 \le p \le \infty$, $1 \le q, q_1 \le \infty$, and $s \ge s_1 > 0$. Assume that either $s > s_1$ or $s = s_1$ and $q \le q_1$. Then $B_p^{s,q} \subset B_p^{s_1,q_1}$.

Proof. Since $l_q(Z^{0+}) \subset l_{q_1}(Z^{0+})$ for $q \le q_1$, we have at once $B_p^{s,q} \subset B_p^{s,q_1}$ for $q \le q_1$. It therefore suffices to treat the case $s > s_1$ and $q > q_1$. Let $v_j = \mathcal{F}^{-1}(\psi_j \hat{v})$ and let r' be the conjugate index to $r = q/q_1$. We then obtain by Hölder's inequality,

$$\|v\|_{B_p^{s_1,q_1}}^{q_1} = \sum_{j=0}^{\infty} 2^{(s_1-s)jq_1} 2^{sjq_1} \|v_j\|_p^{q_1}$$

$$\le \|\{2^{(s_1-s)jq_1}\}\|_{l_{r'}(Z^{0+})} \Big(\sum_{j=0}^{\infty} 2^{sjq} \|v_j\|_p^q \Big)^{1/r} = C\|v\|_{B_p^{s,q}}^{q_1},$$

which proves the theorem.

We now want to compare $B_p^{m,q}$ for m a positive integer with the Sobolev space W_p^m of functions v such that (in the distribution sense) $D^\alpha v \in L_p$ for $|\alpha| \le m$. We set

$$\|v\|_{W_p^m} = \sum_{|\alpha| \le m} \|D^\alpha v\|_p .$$

Theorem 2.2. Let $1 \le p \le \infty$ and let m be a positive integer. Then $B_p^{m,1} \subset W_p^m \subset B_p^{m,\infty}$.

Notice that the norm inequalities corresponding to the above inclusions for $m = 0$ (with $W_p^0 = L_p$) were given in (1.5).

Before giving the proof we introduce an auxiliary partition of unity.

Lemma 2.1. There exist functions $h_k \in C_0^\infty(R^d)$, $k = 1,\dots,d$, with

(2.1) $$\operatorname{supp}(h_k) \subset \{\xi : (4\sqrt{d})^{-1} < |\xi_k| < 4\},$$

and such that

$$\sum_{k=1}^{d} h_k(\xi) = 1 \quad \text{for} \quad \tfrac{1}{2} \le |\xi| \le 2 .$$

Proof. Let $f \in C_0^\infty(R^1)$ be a non-negative function which is positive for $(3\sqrt{d})^{-1} \leq |y| \leq 3$ and with $\text{supp}(f) \subset \{y : (4\sqrt{d})^{-1} < |y| < 4\}$. Then $\sum_{j=1}^{d} f(\xi_j)$ is positive for $1/3 \leq |\xi| \leq 3$. Hence, choosing $g \in C_0^\infty(R^d)$ with $\text{supp}(g) \subset \{\xi : 1/3 < |\xi| < 3\}$ and $g = 1$ for $\frac{1}{2} \leq |\xi| \leq 2$ we find at once that the functions

$$h_k(\xi) = g(\xi)f(\xi_k)/(\sum_{j=1}^{d} f(\xi_j))$$

satisfy the requirements of the lemma.

Proof of Theorem 2.2. Assume first that $v \in W_p^m$. Using the functions h_k, we find by Theorem 1.2.8, for $v_j = \mathcal{F}^{-1}(\psi_j \hat{v})$ with $j > 0$,

$$\|v_j\|_p = \|\mathcal{F}^{-1}(\sum_{k=1}^{d} \frac{h_k(2^{-j}\xi)}{\xi_k^m} \xi_k^m \phi(2^{-j}\xi)\hat{v})\|_p$$

$$\leq 2^{-mj} \sum_{k=1}^{d} M_p(\frac{h_k(\xi)\phi(\xi)}{\xi_k^m})\|\mathcal{F}^{-1}(\xi_k^m \hat{v})\|_p \leq C2^{-mj}\|v\|_{W_p^m} .$$

Since obviously also

$$\|\mathcal{F}^{-1}(\psi_0 \hat{v})\|_p \leq M_p(\psi_0)\|v\|_p \leq C\|v\|_{W_p^m} ,$$

we conclude that

$$\|v\|_{B_p^{m,\infty}} \leq C\|v\|_{W_p^m} ,$$

which proves the first part of the theorem.

Assume next that $v \in B_p^{m,1}$ and let $|\alpha| \leq m$. We have then, in the distribution sense,

$$D^\alpha v = \mathcal{F}^{-1}((i\xi)^\alpha \hat{v}) = \sum_{j=0}^{\infty} \mathcal{F}^{-1}((i\xi)^\alpha \psi_j \hat{v}) .$$

For $j > 0$ we find by Theorem 1.2.8, and since $\psi_j = \psi_j(\phi_{j-1}+\phi_j+\phi_{j+1})$,

$$\|\mathcal{F}^{-1}(\xi^\alpha \psi_j \hat{v})\|_p = \|\mathcal{F}^{-1}(\xi^\alpha \sum_{k=j-1}^{j+1} \phi_k)\hat{v}_j)\|_p$$

$$\leq M_p(\xi^\alpha \sum_{k=j-1}^{j+1} \phi_k)\|v_j\|_p \leq C2^{j|\alpha|}\|v_j\|_p ,$$

and similarly

$$\|\mathcal{F}^{-1}(\xi^\alpha \psi_0 \hat{v})\|_p \leq C\|v_0\|_p ,$$

so that altogether

$$\|v\|_{W_p^m} = \sum_{|\alpha|\leq m} \|\mathcal{F}^{-1}(\xi^\alpha \hat{v})\|_p \leq C \sum_{j=0}^{\infty} 2^{jm}\|v_j\|_p = C\|v\|_{B_p^{m,1}} .$$

This proves the second part of the theorem.

We shall now prove the following density result.

<u>Theorem 2.3</u>. Let $s > 0$, $1 \leq p,q < \infty$. Then C_0^∞ and $\hat{C}_0^\infty$ are dense in $B_p^{s,q}$.

<u>Proof</u>. Let $v \in B_p^{s,q}$ and let $\varepsilon > 0$ be given. Since $q < \infty$ we may choose J such that with $v_j = \mathcal{F}^{-1}(\psi_j \hat{v})$,

$$\|v-v_\varepsilon\|_{B_p^{s,q}} < \varepsilon \quad \text{where} \quad v_\varepsilon = \mathcal{F}^{-1}(\sum_{j=0}^{J} \psi_j \hat{v}) .$$

We have

$$D^\alpha v_\varepsilon = i^{|\alpha|} \mathcal{F}^{-1}((\xi^\alpha \sum_{j=0}^{J} \psi_j)\hat{v}) ,$$

and since $\xi^\alpha \sum_{j=0}^{J} \psi_j \in M_p$ for any α we conclude that $v_\varepsilon \in W_p^m$ for any m. With $m = [s+1]$ we have by Theorems 2.1 and 2.2 that the norm in $B_p^{s,q}$ is dominated by that in W_p^m, and since C_0^∞ and $\hat{C}_0^\infty$ are dense in W_p^m for $p < \infty$, the result follows.

When $p = \infty$ the function $v \equiv 1$ shows that C_0^∞ and $\hat{C}_0^\infty$ are not dense in $B_\infty^{s,q}$. In Section 4 we shall see that (for $d = 1$) $\hat{C}_0^\infty$ is not dense in $B_p^{s,\infty}$ for any s, p.

We shall next show the following Sobolev type embedding result.

<u>Theorem 2.4</u>. Let $1 \leq p < \infty$. Then $B_p^{d/p,1} \subset W_\infty$.

<u>Proof</u>. We have for $j > 0$, and $v \in \hat{C}_0^\infty$,

$$v_j = \mathcal{F}^{-1}(\psi_j \hat{v}) = \sum_{k=j-1}^{j+1} \mathcal{F}^{-1}(\phi_k \hat{v}_j) = \sum_{k=j-1}^{j+1} (\mathcal{F}^{-1}\phi_k) * v_j . \tag{2.2}$$

Since $(\mathcal{F}^{-1}\phi_k)(x) = 2^{kd}\check{\phi}(2^k x)$ we have with p' the conjugate index to p,

$$\|\mathcal{F}^{-1}\phi_k\|_{p'} = 2^{kd}\|\check{\phi}(2^k\cdot)\|_{p'} = 2^{kd/p}\|\check{\phi}\|_{p'} ,$$

and hence, using Hölder's inequality in (2.2),

$$\|v_j\|_\infty \leq C 2^{jd/p}\|v_j\|_p , \quad \text{for } j > 0.$$

Similarly we obtain

$$\|\mathcal{F}^{-1}(\psi_0\hat{v})\|_\infty \leq C\|v_0\|_p ,$$

and hence

$$\|v\|_\infty \leq \sum_{j=0}^{\infty} \|v_j\|_\infty \leq C\|v\|_{B_p^{d/p,1}} .$$

Since $\hat{C}_0^\infty$ is dense in $B_p^{d/p,1}$ (by Theorem 2.3) this proves the theorem.

In Chapter 3 we shall repeatedly use the following corollary.

<u>Corollary 2.1</u>. For any $\nu \geq 0$ there exists a constant C such that if $Av = \mathcal{F}^{-1}(a\hat{v})$ where $a \in C^\infty(R^d)$ is slowly increasing and

$$|a(\xi)| \leq a_0(1 + |\xi|^\nu) \quad \text{for } \xi \in R^d,$$

then for $v \in B_2^{d/2+\nu,1}$,

$$\|Av\|_\infty \le Ca_0\|v\|_{B_2^{d/2+\nu,1}}.$$

Proof. By Theorem 2.4 we have

$$(2.3) \qquad \|Av\|_\infty \le C\|Av\|_{B_2^{d/2,1}}.$$

Since $\mathcal{F}^{-1}(\psi_j\mathcal{F}(Av)) = \mathcal{F}^{-1}(a\psi_j\hat{v})$ we find by Parseval's formula,

$$\|\mathcal{F}^{-1}(\psi_j\mathcal{F}(Av))\|_2 \le \sup_{\xi \in \operatorname{supp}(\psi_j)} |a(\xi)|\, \|\mathcal{F}^{-1}(\psi_j\hat{v})\|_2 \le 2^{1+\nu}a_0 2^{j\nu}\|\mathcal{F}^{-1}(\psi_j\hat{v})\|_2,$$

and hence

$$\|Av\|_{B_2^{d/2,1}} \le 2^{1+\nu}a_0\|v\|_{B_2^{d/2+\nu,1}}.$$

In view of (2.3) this completes the proof.

2.3. An equivalent characterization.

In this section we shall define a norm in $B_p^{s,q}$ for $s > 0$, based on a modulus of continuity in L_p, and show its equivalence to the norms of Section 1.

Let $s > 0$, $1 \le p,q \le \infty$. For $\eta \in R^d$, we set $\Delta_\eta v(x) = v(x+\eta)-v(x)$ and define for $v \in L_p$ and m a positive integer ($m = 1$ or 2 below),

$$\omega_p^m(v;t) = \sup_{|\eta|\le t} \|\Delta_\eta^m v\|_p.$$

We write $s = S+\sigma$ where S is a non-negative integer and $0 < \sigma \le 1$, and set $\bar{\sigma} = 1$ when $\sigma < 1$ (i.e., when s is not an integer) and $\bar{\sigma} = 2$ when $\sigma = 1$ (s integer). We now define for $v \in W_p^S$ (or $v \in L_p$ if $S = 0$),

$$(3.1)\qquad \dot{B}_p^{s,q}(v) = \begin{cases} \sum_{|\alpha|=S} \left(\int_0^\infty (t^{-\sigma}\omega_p^{\bar{\sigma}}(D^\alpha v;t))^q \frac{dt}{t}\right)^{1/q} & \text{for } 1 \le q < \infty , \\ \sum_{|\alpha|=S} \sup_{t>0} t^{-\sigma}\omega_p^{\bar{\sigma}}(D^\alpha v;t) & \text{for } q = \infty . \end{cases}$$

For $p = q = \infty$ the boundedness of the semi-norm $\dot{B}_p^{s,q}(v)$ reduces to the requirement that the derivatives w of v or order S are in Lip_σ if $\sigma < 1$ and if $\sigma = 1$ that they satisfy a Zygmund smoothness condition

$$|w(x+\eta) - 2w(x) + w(x-\eta)| \le C|\eta| .$$

We shall prove that the semi-norm defined in (3.1) is equivalent to the semi-norm $\|\cdot\|_{\dot{B}_p^{s,q}}$ defined in (1.1).

<u>Theorem 3.1</u>. Let $s > 0$, $1 \le p \le \infty$, $1 \le q \le \infty$. Then there exist positive constants c and C such that for $v \in W_p^S$ (or $v \in L_p$ if $S = 0$),

$$c\|v\|_{\dot{B}_p^{s,q}} \le \dot{B}_p^{s,q}(v) \le C\|v\|_{\dot{B}_p^{s,q}} .$$

The proof of Theorem 3.1 will proceed through a sequence of lemmas. Throughout these we assume that $s > 0$ and $1 \le p,q \le \infty$ are fixed, and $s = S+\sigma$, $0 < \sigma \le 1$, S integer, Further, v denotes a function in W_p^S (or L_p if $S = 0$).

In our first lemma we will estimate the integral (supremum) in the definition (3.1) from above and below by the l_q norm of an associated sequence.

<u>Lemma 3.1</u>. We have

$$(\log 2)^{1/q} 2^{-\sigma} \sum_{|\alpha|=S} \|\{2^{\sigma j}\omega_p^{\bar{\sigma}}(D^\alpha v;2^{-j})\}_{j\in Z}\|_{l_q} \le \dot{B}_p^{s,q}(v)$$

$$\le (\log 2)^{1/q} 2^{\sigma} \sum_{|\alpha|=S} \|\{2^{\sigma j}\omega_p^{\bar{\sigma}}(D^\alpha v;2^{-j})\}_{j\in Z}\|_{l_q} .$$

Proof. Assume that $q < \infty$; the case $q = \infty$ is similar. Put for $|\alpha| = S$,

$$B_\alpha = \int_0^\infty (t^{-\sigma}\bar{\omega}_p^{\sigma}(D^\alpha v;t))^q \frac{dt}{t} .$$

We make the transformation of variables $t = 2^x$ and obtain

$$B_\alpha = \log 2 \int_{-\infty}^{\infty} (2^{-\sigma x}\bar{\omega}_p^{\sigma}(D^\alpha v;2^x))^q dx .$$

Notice that $2^{-\sigma qx}$ is a decreasing function of x, and that $\bar{\omega}_p^{\sigma}(v;2^x)^q$ is increasing. Hence for $-j \le x \le -j+1$,

$$(2^{\sigma(j-1)}\bar{\omega}_p^{\sigma}(D^\alpha v;2^{-j}))^q \le (2^{-\sigma x}\bar{\omega}_p^{\sigma}(D^\alpha v;2^x))^q \le (2^{\sigma j}\bar{\omega}_p^{\sigma}(D^\alpha v;2^{-j+1}))^q .$$

Writing

$$B_\alpha = \log 2 \sum_{j=-\infty}^{\infty} \int_{-j}^{-j+1} (2^{-\sigma x}\bar{\omega}_p^{\sigma}(D^\alpha v;2^x))^q dx ,$$

we obtain

$$(\log 2)2^{-\sigma q} \sum_{j=-\infty}^{\infty} (2^{\sigma j}\bar{\omega}_p^{\sigma}(D^\alpha v;2^{-j}))^q \le B_\alpha \le (\log 2)2^{\sigma q} \sum_{j=-\infty}^{\infty} (2^{\sigma j}\bar{\omega}_p^{\sigma}(D^\alpha v;2^{-j}))^q .$$

This proves the lemma.

In the proof of the next lemma we shall use the functions h_k defined in Lemma 2.1.

Lemma 3.2. There exists a constant C such that for $j \in Z$,

$$2^{Sj}\|\mathcal{F}^{-1}(\phi_j\hat{v})\|_p \le C2^{\sigma j} \sum_{|\alpha|=S} \bar{\omega}_p^{\sigma}(D^\alpha v;2^{-j}).$$

Proof. Letting $h_{kj}(\xi) = h_k(2^{-j}\xi)$ we obtain for $v_j = \mathcal{F}^{-1}(\phi_j\hat{v})$,

$$v_j = \sum_{k=1}^{d} \mathcal{F}^{-1}(h_{kj}\phi_j\hat{v}).$$

Setting $t_{kj}(\xi) = \exp(i2^{-j}\xi_k) - 1$ we have with e_k the kth unit vector,

$$\Delta^{\bar{\sigma}}_{2^{-j}e_k} \left(\frac{\partial}{\partial x_k}\right)^S v = i^S \mathcal{F}^{-1}(t^{\bar{\sigma}}_{kj}\xi^S_k \hat{v}) .$$

Since by (2.1), t_{kj} and ξ_k are non-zero on $\operatorname{supp}(h_{kj})$, we can define $H_{kj} \in C^\infty_0(R^d)$ by

$$H_{kj}(\xi) = h_{kj}(\xi)\phi_j(\xi)/(t^{\bar{\sigma}}_{kj}(\xi) \cdot 2^{-jS}\xi^S_k) = H_{k0}(2^{-j}\xi) ,$$

and obtain, using also Theorem 1.2.8,

$$\|v_j\|_p \le 2^{-Sj} \sum_{k=1}^{d} M_p(H_{kj})\|\mathcal{F}^{-1}(t^{\bar{\sigma}}_{kj}\xi^S_k\hat{v})\|_p$$

$$= 2^{-Sj} \sum_{k=1}^{d} M_p(H_{k0})\|\Delta^{\bar{\sigma}}_{2^{-j}e_k}\left(\frac{\partial}{\partial x_k}\right)^S v\|_p \le C2^{-Sj} \sum_{|\alpha|=S} \omega^{\bar{\sigma}}_p(D^\alpha v;2^{-j}) .$$

Multiplying each side by 2^{Sj}, this proves the lemma.

The next lemma will be used in the proof of Lemma 3.4, which is the "converse" of Lemma 3.2.

Lemma 3.3. There exists a constant C such that for $l \in Z$, $\eta \in R^d$ and $|\alpha| = S$,

$$(3.2) \qquad M_p((e^{i\langle\eta,\xi\rangle} - 1)^{\bar{\sigma}}\xi^\alpha\phi_l) \le C2^{Sl}\min((|\eta|2^l)^{\bar{\sigma}},1).$$

Proof. Using Theorems 1.2.7 and 1.2.8 we find that

$$M = M_p((e^{i\langle\eta,\xi\rangle} - 1)^{\bar{\sigma}}\xi^\alpha\phi_l) \le M^{(1)}_p((e^{iy} - 1)^{\bar{\sigma}})M_p(\xi^\alpha\phi_l) \le C2^{Sl} .$$

To complete the proof of (3.2) notice that for $d = 1$, the Carlson-Beurling inequality shows that $f(y) = (e^{iy} - 1)/y \in M_\infty \subset M_p$, and hence we have by Theorem 1.2.8,

$$M \le M_p(f(\langle\eta,\xi\rangle))^{\bar{\sigma}}M_p(\langle\eta,\xi\rangle^{\bar{\sigma}}\phi_l) \sum_{k=l-1}^{l+1} M_p(\xi^\alpha\phi_k)$$

$$\le CM^{(1)}_p(f)^{\bar{\sigma}}M_p(\langle\eta 2^l,\xi\rangle^{\bar{\sigma}}\phi_0)2^{Sl} \le C(|\eta|2^l)^{\bar{\sigma}}2^{Sl} .$$

This completes the proof of the lemma.

Lemma 3.4. There exists a constant C such that for $j \in Z$ and $|\alpha| = S$,

$$\omega_p^{\bar{\sigma}}(D^\alpha v; 2^{-j}) \leq C \sum_{k=-\infty}^{\infty} 2^{Sk} \min(2^{(k-j)\bar{\sigma}}, 1) \|\mathcal{F}^{-1}(\phi_k \hat{v})\|_p .$$

Proof. Let $v_k = \mathcal{F}^{-1}(\phi_k \hat{v})$. We have

$$\Delta_\eta^{\bar{\sigma}} D^\alpha v = \sum_{k=-\infty}^{\infty} \sum_{l=k-1}^{k+1} \mathcal{F}^{-1}((e^{i<\eta,\xi>} - 1)^{\bar{\sigma}} (i\xi)^\alpha \phi_l \hat{v}_k) ,$$

so that

$$\|\Delta_\eta^{\bar{\sigma}} D^\alpha v\|_p \leq \sum_{k=-\infty}^{\infty} \sum_{l=k-1}^{k+1} M_p((e^{i<\eta,\xi>} - 1)^{\bar{\sigma}} \xi^\alpha \phi_l) \|v_k\|_p .$$

Using Lemma 3.3 and taking supremum over $|\eta| \leq 2^{-j}$ proves the lemma.

We can now prove Theorem 3.1.

Proof of Theorem 3.1. Let $v_j = \mathcal{F}^{-1}(\phi_j \hat{v})$. We have by Lemmas 3.2 and 3.1,

$$\|v\|_{\dot{B}_p^{s,q}} = \|\{2^{sj}\|v_j\|_p\}\|_{l_q} \leq C \sum_{|\alpha|=S} \|\{2^{\sigma j} \omega_p^{\bar{\sigma}}(D^\alpha v; 2^{-j})\}\|_{l_q} \leq C\dot{B}_p^{s,q}(v) .$$

This proves one half of the theorem.

In order to prove the converse inequality, recall the discrete Young's inequality: for sequences $c^{(1)}$ and $c^{(2)}$,

$$\|\{\sum_{k=-\infty}^{\infty} c^{(1)}_{\cdot -k} c^{(2)}_k\}\|_{l_q} \leq \|c^{(1)}\|_{l_1} \|c^{(2)}\|_{l_q} .$$

Using Lemmas 3.1 and 3.4 we have

$$\dot{B}_p^{s,q}(v) \leq C \sum_{|\alpha|=S} \|\{2^{\sigma j} \omega_p^{\bar{\sigma}}(D^\alpha v; 2^{-j})\}\|_{l_q}$$

$$\leq C \|\{\sum_{k=-\infty}^{\infty} 2^{\sigma(j-k)} \min(2^{-(j-k)\bar{\sigma}}, 1) 2^{sk} \|v_k\|_p\}\|_{l_q}$$

$$\leq C \|\{2^{\sigma j} \min(2^{-j\bar{\sigma}}, 1)\}\|_{l_1} \|\{2^{sj}\|v_j\|_p\}\|_{l_q} .$$

Since $\bar{\sigma} > \sigma > 0$ we have

$$\|\{2^{\sigma j}\min(2^{-j\bar{\sigma}},1)\}\|_{l_1} = \sum_{j=-\infty}^{0} 2^{\sigma j} + \sum_{j=1}^{\infty} 2^{(\sigma-\bar{\sigma})j} < \infty ,$$

and hence

$$\dot{B}_p^{s,q}(v) \leq C\|v\|_{\dot{B}_p^{s,q}} .$$

This completes the proof of the theorem.

Notice that the constant c determined in the proof can be chosen uniformly for s in any bounded interval, whereas C tends to infinity as s tends to an integer from below.

It follows from Theorems 2.1 and 2.2 that if $v \in B_p^{s,q}$ with $s > 0$, then $v \in W_p^S$, and hence by Theorems 1.1 and 3.1 that $\|\cdot\|_p + \dot{B}_p^{s,q}(\cdot)$ is a norm in $B_p^{s,q}$, equivalent to $\|\cdot\|_{B_p^{s,q}}$.

2.4. Two examples.

In this section we shall exhibit two special functions of one variable which will be used later to discuss sharpness of certain estimates.

Example I. Let G be a non-identically vanishing function with $\hat{G} \in C_0^\infty(0,1)$ and set, for $\tau > 0$,

$$G_\tau(x) = \left(\sum_{j=1}^{\infty} \exp(i2^j x)2^{-\tau j}\right)G(x).$$

Notice that since the sum is absolutely convergent, $G_\tau \in W_p$ for $1 \leq p \leq \infty$. We have the following:

Proposition 4.1. Let $1 \leq p \leq \infty$. Then $G_\tau \in B_p^{s,q}$ if and only if $s < \tau$ or $s = \tau$, $q = \infty$.

Proof. We have

$$\hat{G}_\tau(\xi) = \sum_{j=1}^{\infty} 2^{-\tau j}\hat{G}(\xi-2^j).$$

The support of the jth term in this sum is contained in the interval $(2^j, 2^j+1)$ and is hence, for large j, disjoint from the support of ψ_l for $l \neq j$. Since then $\psi_j = 1$ on $(2^j, 2^j+1)$ we obtain for such j,

$$\mathcal{F}^{-1}(\psi_j \hat{G}_\tau)(x) = 2^{-\tau j}\exp(i2^j x)G(x),$$

and hence

$$\|\mathcal{F}^{-1}(\psi_j \hat{G}_\tau)\|_p = 2^{-\tau j}\|G\|_p, \tag{4.1}$$

from which the proposition follows by (1.3).

It can be proved that for τ a positive integer, $D^{\tau-1}G_\tau$ is continuous, but $D^\tau G_\tau$ is non-existent a.e.

The function G_τ may be used to show that $\hat{C}_0^\infty$ is not dense in $B_p^{\tau,\infty}$ for any $1 \leq p \leq \infty$ (for $d = 1$). For, if v is any function in $\hat{C}_0^\infty$ we have by (4.1) for large j,

$$\|\mathcal{F}^{-1}(\psi_j(\hat{v} - \hat{G}_\tau))\|_p = 2^{-\tau j}\|G\|_p$$

so that

$$\|v - G_\tau\|_{B_p^{\tau,\infty}} \geq \|G\|_p > 0 .$$

Example II. Let $H \in C_0^\infty(-1,1)$ be a non-negative function with $H(0) > 0$ and set

$$H_\tau(x) = x_+^\tau H(x),$$

where

$$x_+^\tau = \begin{cases} x^\tau & \text{for } x > 0, \\ 0 & \text{for } x \leq 0 . \end{cases}$$

Then H_τ has compact support and is smooth outside the origin. We shall prove the following:

Proposition 4.2. Let $1 \leq p \leq \infty$ and $\tau + 1/p > 0$. Then $H_\tau \in B_p^{s,q}$ if and only if $s < \tau + 1/p$ or $s = \tau + 1/p$, $q = \infty$.

Proof. By Theorem 2.1 it suffices to prove that with $s = \tau + 1/p$, H_τ belongs to $B_p^{s,\infty}$ but not to $B_p^{s,q}$ for $q < \infty$. We write $s = S + \sigma$ where S is a non-negative integer and $0 < \sigma \leq 1$. Then $S < \tau + 1/p$ and since $D^S H_\tau(x) = O(|x|^{\tau-S})$ as $x \to 0$, we have $H_\tau \in W_p^S$ (or L_p if $S = 0$). By the second remark following Theorem 3.1 the claim is therefore that $\dot{B}_p^{s,q}(H_\tau)$ is finite for $q = \infty$ but infinite otherwise. Since $\dot{B}_p^{s,q}(H_\tau) = \dot{B}_p^{\sigma,q}(D^S H_\tau)$ and since $D^S H_\tau$ is a function of the same form as H_τ but with τ replaced by $\tau - S$, we find that it is now sufficient to consider the case $0 < s = \tau + 1/p \leq 1$. We shall carry out the proof only for $p < \infty$; the case $p = \infty$ is analogous.

We start with the case $s < 1$ and consider first points for which $|x| \geq 2|\eta|$ so that x and $x+\eta$ have the same sign and $|x+\eta| \geq |x|/2$. We then obtain, by the mean value theorem,

$$|\Delta_\eta H_\tau(x)| \leq C|\eta| \sup_{|y| \geq |x|/2} |H_\tau'(y)| \leq C|\eta| x_+^{\tau-1} .$$

Using for $|x| < 2|\eta|$ the obvious estimate

$$|\Delta_\eta H_\tau(x)| \leq H_\tau(x) + H_\tau(x+\eta) \leq C[x_+^\tau + (x+\eta)_+^\tau] ,$$

we hence obtain, since $(\tau-1)p + 1 = -p(1-s) < 0$ and $\tau p + 1 = ps > 0$,

$$\|\Delta_\eta H_\tau\|_p \leq C|\eta| \Big(\int_{2|\eta|}^{\infty} x^{(\tau-1)p} dx \Big)^{1/p} + C\Big(\int_0^{3|\eta|} x^{\tau p} dx \Big)^{1/p} \leq C|\eta|^s .$$

It follows at once that

$$\omega_p^1(H_\tau; t) \leq ct^s ,$$

so that $\dot{B}_p^{s,\infty}(H_\tau)$ is finite. On the other hand, since for x, t positive and small,

$$|\Delta_t H_\tau(x)| \geq ct(x+t)^{\tau-1} \quad \text{with } c > 0,$$

we find for t small,

$$\omega_p^1(H_\tau;t) \geq \|\Delta_t H_\tau\|_p \geq ct(\int_0^t (x+t)^{(\tau-1)p}dx)^{1/p} \geq ct^s .$$

Hence $\dot{B}_p^{s,q}(H_\tau) = \infty$ for $q < \infty$.

For $s = 1$, we have similarly

$$|\Delta_\eta^2 H_\tau(x)| \leq \begin{cases} C|\eta|^2 x_+^{\tau-2}, & \text{for } |x| \geq 3|\eta| , \\ C[x_+^\tau + (x+\eta)_+^\tau + (x+2\eta)_+^\tau], & \text{for } |x| < 3|\eta| , \end{cases}$$

and hence, since now $(\tau-2)p+1 = -p$ and $\tau p+1 = p$,

$$\|\Delta_\eta^2 H_\tau\|_p \leq C|\eta|^2(\int_{3|\eta|}^\infty x^{(\tau-2)p}dx)^{1/p} + C(\int_0^{5|\eta|} x^{\tau p}dx)^{1/p} \leq C|\eta| ,$$

so that we may conclude that $\dot{B}_p^{1,\infty}(H_\tau)$ is finite. Here we have for x,t positive and small,

$$|\Delta_t H_\tau(x)| \geq ct^2(x+t)^{\tau-2} \quad \text{with } c > 0,$$

so that $\omega_p^2(H_\tau;t) \geq ct$ and hence $\dot{B}_p^{1,q}(H_\tau) = \infty$ for $q < \infty$. This concludes the proof of the proposition.

Notice that in contrast to the function G_τ of Example I, the function H_τ, which has its non-smoothness concentrated to one point, permits higher values of s, the smaller p is, so that for instance $H_{1/2}$, which belongs to $B_\infty^{1/2,\infty} = \mathrm{Lip}_{1/2}$ but not to $B_\infty^{s,\infty}$ for $s > \frac{1}{2}$, does belong to $B_1^{3/2,\infty}$.

2.5. An interpolation property.

In this section we shall show that in a certain sense the Besov space $B_p^{s,\infty}$ interpolates a pair of such spaces with larger and smaller values of s. For the definition of the space $B_p^{0,1}$, recall the remark following Theorem 1.1.

Theorem 5.1. Let $1 \le p \le \infty$, and $s_1 > s_0 \ge 0$, and let V be a normed linear space with norm $\|\cdot\|_V$. Then for s with $s_1 > s > s_0$ there is a constant C such that if T is a bounded linear operator from $B_p^{s_0,1}$ into V and

$$\|Tv\|_V \le N_l \|v\|_{B_p^{s_l,1}}, \quad \text{for } v \in B_p^{s_l,1}, \quad l = 0,1,$$

then with $\theta = (s-s_0)/(s_1-s_0)$,

$$(5.1) \qquad \|Tv\|_V \le C N_0^{1-\theta} N_1^{\theta} \|v\|_{B_p^{s,\infty}}, \text{ for } v \in B_p^{s,\infty}.$$

Proof. We have for $v_j = \mathcal{F}^{-1}(\psi_j \hat{v})$, $j \ge 0$,

$$\|Tv_j\|_V \le \min(N_0 \|v_j\|_{B_p^{s_0,1}}, N_1 \|v_j\|_{B_p^{s_1,1}})$$

$$\le C \min(N_0 2^{js_0}, N_1 2^{js_1}) \|v_j\|_p \le C 2^{-sj} \min(N_0 2^{js_0}, N_1 2^{js_1}) \|v\|_{B_p^{s,\infty}}.$$

Let now j_0 be the smallest non-negative integer such that

$$2^{j_0(s_1-s_0)} > \frac{N_0}{N_1}.$$

We then obtain

$$(5.2) \qquad \begin{aligned} \sum_{j=0}^{\infty} 2^{-sj} \min(N_0 2^{js_0}, N_1 2^{js_1}) &= N_0 \sum_{j \ge j_0} 2^{-j(s-s_0)} + N_1 \sum_{j<j_0} 2^{j(s_1-s)} \\ &\le C_s [N_0 2^{-j_0(s-s_0)} + N_1 2^{j_0(s_1-s)}] \le C_s N_0^{1-\theta} N_1^{\theta}, \end{aligned}$$

and hence

$$\|Tv\|_V \le \sum_{j=0}^{\infty} \|Tv_j\|_V \le C_s N_0^{1-\theta} N_1^{\theta} \|v\|_{B_p^{s,\infty}},$$

which proves the theorem.

Notice that the constants C_s in (5.2) tend to infinity as θ tends to 0 or 1. It can be proved that if we replace $B_p^{s,\infty}$ by $B_p^{s,1}$ in (5.1), then the constant here can be estimated in terms of s_0 and s_1 alone.

Using the previously proven embedding results we may conclude the following:

Corollary 5.1. Let $1 \le p \le \infty$, and $s_1 > s_0 \ge 0$ and let for $l = 0,1$, $B_l = B_p^{s_l,q_l}$ $(1 \le q_l \le \infty)$, $W_p^{s_l}$ $(Z \ni s_l > 0)$ or L_p $(s_0 = 0)$. Then for s with $s_0 < s < s_1$ there is a constant C such that if T is a bounded linear operator from B_0 into a normed linear space V and

$$\|Tv\|_V \le N_l \|v\|_{B_l} \quad \text{for } v \in B_l, \ l = 0,1,$$

then with $\theta = (s-s_0)/(s_1-s_0)$,

$$\|Tv\|_V \le CN_0^{1-\theta} N_1^{\theta} \|v\|_{B_p^{s,\infty}} \quad \text{for } v \in B_p^{s,\infty}.$$

Proof. Since by Theorem 2.1, Theorem 2.2 or (1.5) as the case may be,

$$\|Tv\|_V \le N_l \|v\|_{B_l} \le CN_l \|v\|_{B_p^{s_l,1}},$$

the result follows from Theorem 5.1.

2.6. Two special operator estimates.

For later reference we shall formulate here two simple estimates for operators of the form $Av = \mathcal{F}^{-1}(a\hat{v})$ from $B_p^{s,q}$ into L_p. We first have the following.

Lemma 6.1. Let $1 \le p,q \le \infty$ and $s > 0$. Let $a \in C^\infty$ be slowly increasing, and assume that with q' the conjugate exponent to q,

$$M_p^{s,q'}(a) = \|\{2^{-sj} M_p(\psi_j a)\}\|_{l_{q'}(Z^{0+})} < \infty .$$

Then for $v \in B_p^{s,q}$ we have $Av \equiv \mathcal{F}^{-1}(a\hat{v}) \in L_p$ and

$$\|Av\|_p \le 3 \cdot 2^s M_p^{s,q'}(a) \|v\|_{B_p^{s,q}} .$$

Proof. Let $v_j = \mathcal{F}^{-1}(\psi_j \hat{v})$, $j \geq -1$, with $\psi_{-1} \equiv 0$. We have

(6.1) $$Av = \sum_{j=0}^{\infty} \mathcal{F}^{-1}\Big(\Big(\sum_{k=j-1}^{j+1} \psi_k\Big) a \mathcal{F} v_j\Big) .$$

Now, for $|k-j| \leq 1$,

$$\| \mathcal{F}^{-1}(\psi_k \, a \mathcal{F} v_j)\|_p \leq M_p(\psi_k a)\|v_j\|_p \leq 2^s 2^{-sk} M_p(\psi_k a) 2^{sj} \|v_j\|_p .$$

Hence, applying Hölder's inequality in (6.1) we find that $Av \in L_p$ if $v \in B_p^{s,q}$ and that the desired inequality holds. This proves the lemma.

The following estimate using the seminorm instead of the norm in $B_p^{s,q}$ is proved analogously.

Lemma 6.2. Let $1 \leq p,q \leq \infty$ and $s > 0$. Let $a \in C^\infty$ be slowly increasing, and assume that with q' the conjugate exponent to q,

$$\dot{M}_p^{s,q'}(a) = \| \{2^{-sj} M_p(\phi_j a)\} \|_{l_{q'}(Z)} < \infty .$$

Then for $v \in B_p^{s,q}$ we have $Av \equiv \mathcal{F}^{-1}(a\hat{v}) \in L_p$ and

$$\|Av\|_p \leq 3 \cdot 2^s \dot{M}_p^{s,q'}(a) \|v\|_{\dot{B}_p^{s,q}} .$$

References.

Our main source for the theory of Besov spaces has been [5]; see also [3] and references for related material. The Sobolev type embedding result of Theorem 2.4 was proved in [6] in a somewhat sharper form under the name of Bernstein's theorem. The proof in Section 3 of the equivalence between the definition in Section 1 and Besov's original definition [2] follows [4]. The function G_τ of Section 4 was studied in [1, p. 265].

1. N.K. Bari, A Treatise on Trigonometric Series, vol. 2, McMillan, New York 1964.

2. O.V. Besov, Investigation of a family of function spaces in connection with theorems of embedding and extension (Russian), Trudy. Mat. Inst. Steklov. 60 (1961), 42-81 = Amer. Math. Soc. Transl. (2) 40 (1964), 85-126.

3. P.L. Butzer and H. Berens, Semi-Groups of Operators and Approximation, Springer, Berlin 1967.

4. B. Grevholm, On the structure of the spaces $\mathcal{L}_k^{p,\lambda}$, Math. Scand. 26 (1970), 241-254.

5. J. Peetre, Reflexions about Besov spaces (Swedish), Department of Mathematics, University of Lund, Lund 1966.

6. J. Peetre, Applications de la theorie des espaces d'interpolation dans l'analyse harmonique, Ricerche Mat. 15 (1966), 1-36.

CHAPTER 3. INITIAL VALUE PROBLEMS AND DIFFERENCE OPERATORS.

In this chapter, which has the character of an introduction to the rest of these notes, we shall consider initial value problems for first-order-in-time constant coefficient (scalar) partial differential equations and finite difference approximations to such problems. By application of Parseval's formula we shall obtain convergence estimates in L_2 for a large class of equations and difference methods, and also certain such estimates in the maximum norm which can be obtained similarly by means of Sobolev's inequality. In later chapters we shall employ the techniques developed in Chapters 1 and 2 to investigate in more detail the convergence in the maximum norm (and in L_p for $p \neq 2$) for parabolic equations, first order hyperbolic equations, and the Schrödinger equation.

3.1. Well posed initial value problems.

Consider the initial value problem for $u = u(x,t)$,

$$\begin{aligned} &\frac{\partial u}{\partial t} = Pu = \sum_{|\alpha| \leq M} p_\alpha D^\alpha u, \quad \text{for} \quad x \in R^d,\ t > 0, \\ &u(x,0) = v(x). \end{aligned} \tag{1.1}$$

Here the p_α are complex numbers and the integer M is the order of the operator P.

The characteristic polynomial or symbol $\hat{P}$ of P is defined by

$$\hat{P}(\xi) = \sum_{|\alpha| \leq M} p_\alpha (i\xi)^\alpha ,$$

so that for $u \in S'$ we have $Pu = \mathcal{F}^{-1}(\hat{P}\hat{u})$.

Suppose that (1.1) has a solution $u(t) = u(x,t) \in S'$. Taking Fourier transforms with respect to x, this leads to the following initial value problem for an ordinary differential equation for $\hat{u}(\xi,t)$,

$$\frac{d\hat{u}}{dt} = \hat{P}(\xi)\hat{u}, \ t > 0,$$

$$\hat{u}(\cdot,0) = \hat{v},$$

which may formally be integrated to yield

$$\hat{u}(\xi,t) = \exp(t\hat{P}(\xi))\hat{v}(\xi).$$

On the other hand, at least for $v \in \hat{C}_0^\infty$ we have with $\hat{u}$ thus defined that $u = \mathcal{F}^{-1}\hat{u}$ is a smooth classical solution of (1.1). This motivates the following definition: For $t \geq 0$ the solution operator $E(t)$ of the problem (1.1) is

$$E(t)v = \mathcal{F}^{-1}(\exp(t\hat{P})\hat{v}), \ \text{for} \ v \in \hat{C}_0^\infty.$$

We say that (1.1) is well posed in L_p if for any $T > 0$ there exists a constant C such that

$$(1.2) \qquad \|E(t)v\|_p \leq C\|v\|_p, \ \text{for} \ t \leq T, \ v \in \hat{C}_0^\infty.$$

(Here and in the sequel we only consider non-negative t.)

In fact, it suffices to demand (1.2) for one value of $T > 0$. For, assuming (1.2) we may write an arbitrary t as $t = \tau+mT$, with $0 \leq \tau < T$ and m integer, and it follows that

$$\|E(t)v\|_p = \|E(\tau)E^m(T)v\|_p \leq C^{m+1}\|v\|_p$$

so that

$$(1.3) \qquad \|E(t)v\|_p \leq C_1 e^{C_2 t}\|v\|_p, \ \text{for} \ t \geq 0, \ v \in \hat{C}_0^\infty.$$

The property (1.2) can also be expressed in Fourier multiplier terminology:

<u>Proposition 1.1</u>. The initial value problem (1.1) is well posed in L_p if and only if $\exp(t\hat{P}) \in M_p$ for $t \geq 0$, and for any $T > 0$ there exists a constant C such that

$$M_p(\exp(t\hat{P})) \leq C, \ \text{for} \ t \leq T.$$

In particular, (1.1) is well posed in L_2 if and only if for some constant C,

$$\operatorname{Re} \hat{P}(\xi) \leq C, \ \xi \in R^d.$$

Proof. The first part is obvious from the definitions. The second follows since

$$M_2(\exp(t\hat{P})) = \sup_\xi |\exp(t\hat{P}(\xi))| = \exp(t \sup_\xi \operatorname{Re} \hat{P}(\xi)),$$

which proves the proposition.

We notice that if (1.1) is well posed in L_p, then by Proposition 1.1 and Theorem 1.2.3 the solution operator $E(t)$ admits a bounded extension to all of L_p which we shall for simplicity also denote by $E(t)$.

As examples, consider the following three simple equations, which will be discussed in more detail in subsequent chapters, namely the heat equation

$$\frac{\partial u}{\partial t} = \Delta u = \sum_{j=1}^{d} \frac{\partial^2 u}{\partial x_j^2},$$

the one-dimensional first order hyperbolic equation

$$\frac{\partial u}{\partial t} = \frac{\partial u}{\partial x},$$

and the Schrödinger equation

$$\frac{\partial u}{\partial t} = i\Delta u .$$

For the heat equation we have $\hat{P}(\xi) = -|\xi|^2$, so that for each $t > 0$, $\exp(t\hat{P}(\xi)) = \exp(-t|\xi|^2)$ belongs to S, and hence, as is easily seen, belongs to M_p for $1 \leq p \leq \infty$. We also find by the change of variables $\xi \to t^{-1/2}\xi$, using Theorem 1.2.8, that

$$M_p(\exp(t\hat{P}(\xi))) = M_p(\exp(-|\xi|^2)) = \text{constant}.$$

For the hyperbolic equation above, $\hat{P}(\xi) = i\xi$ and

$$M_p(\exp(t\hat{P}(\xi))) = M_p(e^{it\xi}) = 1 ;$$

in this case the solution operator is just the translation operator $E(t)v(x) = v(x+t)$. For the Schrödinger equation we have $\hat{P}(\xi) = -i|\xi|^2$, and $\exp(-it|\xi|^2)$ is in M_2 but by Corollary 1.5.3 not in M_p for $p \neq 2$. We conclude that the initial value problems for the first two equations are well posed in L_p for $1 \leq p \leq \infty$, whereas the initial value problem for the Schrödinger equation is well posed only in L_2.

In all three examples the relevant norms are independent of t, since the characteristic polynomials are homogeneous in ξ. An example exhibiting the growth in t indicated in (1.3) is given by

$$\frac{\partial u}{\partial t} = \frac{\partial u}{\partial x} + cu, \ c > 0,$$

in which case

$$M_p(\exp(t\hat{P}(\xi)) = M_p(e^{t(i\xi+c)}) = e^{ct} .$$

For the backward heat equation

$$\frac{\partial u}{\partial t} = -\Delta u ,$$

we have $\operatorname{Re} \hat{P}(\xi) = |\xi|^2$, which is unbounded so that the corresponding initial value problem is ill-posed even in L_2.

For equations like the Schrödinger equation for which we have well-posedness only in L_2, it is still possible to derive maximum norm estimates for the solution under appropriate regularity assumptions on the initial data. The following is one result in this direction.

<u>Theorem 1.1</u>. Let the initial value problem (1.1) be well posed in L_2. Then for each $T > 0$ there is a constant C such that for $v \in B_2^{d/2,1}$,

$$\|E(t)v\|_\infty \leq C\|v\|_{B_2^{d/2,1}} , \quad \text{for} \quad t \leq T.$$

<u>Proof</u>. This follows immediately from Corollary 2.2.1 since $\exp(t\hat{P})$ is uniformly bounded for $0 \leq t \leq T$.

3.2. Finite difference operators and stability.

In this section we shall present preliminary material on semi-discrete and completely discrete approximations to the initial value problem (1.1).

Let h be a (small) positive number and let for $\beta \in Z^d$, T_h^β denote the translation operator defined by $T_h^\beta v(x) = v(x+h\beta)$. We shall consider approximations P_h of P of the form

$$(2.1) \qquad P_h v = h^{-M} \sum_\beta p_\beta(h) T_h^\beta v ,$$

where the $p_\beta(h)$ are polynomials in h, and the summation is over a finite set in Z^d. Such an approximation is said to be consistent with P if for any $x \in R^d$ and any function v which is smooth in a neighborhood of x,

$$(2.2) \qquad P_h v(x) = Pv(x) + o(1), \text{ as } h \to 0.$$

Consistent approximations can be obtained, for instance, by replacing the derivatives D^α in P by the appropriate symmetric difference quotients,

$$\partial_h^\alpha = \partial_1^{\alpha_1} \cdots \partial_d^{\alpha_d} \quad \text{where} \quad \partial_j v(x) = \frac{v(x+he_j) - v(x-he_j)}{2h} .$$

Here e_j denotes the jth unit vector.

Notice that for P_h defined by (2.1) we have $P_h v = \mathcal{F}^{-1}(\hat{P}_h \hat{v})$, where

$$\hat{P}_h(\xi) = h^{-M} \sum_\beta p_\beta(h) e^{i\langle\xi, h\beta\rangle} .$$

For the particular operator $P_h = \sum_{|\alpha| \le M} p_\alpha \partial_h^\alpha$ we then obtain

$$\hat{P}_h(\xi) = \hat{P}(h^{-1}\sin(h\xi)), \text{ where } \sin \xi = (\sin \xi_1, \ldots, \sin \xi_d).$$

The consistency condition (2.2) is equivalent to a finite number of linear equations between the coefficients of P_h and P. Using this we shall now see that the consistency can be expressed in terms of the symbols $\hat{P}_h$ and $\hat{P}$.

Proposition 2.1. The operator P_h is consistent with P if and only if

$$(2.3) \qquad h^M \hat{P}_h(h^{-1}\xi) = h^M \hat{P}(h^{-1}\xi) + o(h^M + |\xi|^M), \text{ as } h, \xi \to 0 .$$

Proof. Let in (2.1), $p_\beta(h) = \sum_j p_{\beta j} h^j$. Developing both sides of (2.2) in Taylor series around x, we obtain after multiplication by h^M,

$$\sum_{j+|\alpha|\le M} h^{j+|\alpha|} \sum_\beta \beta^\alpha p_{\beta j} \frac{1}{\alpha!} D^\alpha v(x) = h^M \sum_\alpha p_\alpha D^\alpha v(x) + o(h^M), \text{ as } h \to 0.$$

Since the $D^\alpha v(x)$ are arbitrary we conclude that (2.2) holds if and only if

$$(2.4) \qquad \frac{1}{\alpha!} \sum_\beta \beta^\alpha p_{\beta j} = \begin{cases} 0 & \text{for } j+|\alpha| < M, \\ p_\alpha & \text{for } j+|\alpha| = M. \end{cases}$$

On the other hand, as h and ξ tend to zero,

$$h^M \hat{P}_h(h^{-1}\xi) = \sum_\beta p_\beta(h) e^{i<\xi,\beta>} = \sum_{\beta,s} p_\beta(h) \frac{i^s}{s!} <\beta,\xi>^s$$

$$= \sum_{j+|\alpha|\le M} h^j \sum_{\beta,\alpha} \beta^\alpha p_{\beta j} \frac{i^{|\alpha|}}{\alpha!} \xi^\alpha + o(h^M + |\xi|^M),$$

and

$$h^M \hat{P}(h^{-1}\xi) = \sum_{|\alpha|\le M} h^{M-|\alpha|} p_\alpha (i\xi)^\alpha .$$

Hence (2.3) holds if and only if (2.4) is valid, which completes the proof.

As a result of Proposition 2.1 we see that (2.3) could have been used instead of (2.2) as a definition. In analogous situations below we shall find it convenient to express the definitions directly in terms of symbols.

Consider now first the initial value problem obtained by replacing P in (1.1) by a consistent finite difference operator P_h,

$$(2.5) \qquad \begin{aligned} \frac{\partial u_h}{\partial t} &= P_h u, \ t > 0, \\ u_h(\cdot,0) &= v . \end{aligned}$$

As in Section 1, taking Fourier transforms with respect to x, this leads to an initial value problem for an ordinary differential equation,

$$\frac{d\hat{u}_h}{dt} = \hat{P}_h \hat{u}_h, \quad t > 0,$$

$$\hat{u}_h(\cdot,0) = \hat{v},$$

which has the solution

$$\hat{u}_h(\xi,t) = \exp(t\hat{P}_h(\xi))\hat{v}(\xi) .$$

We may therefore define the solution operator of the semi-discrete problem by

$$E_h(t)v = \mathcal{F}^{-1}(\exp(t\hat{P}_h)\hat{v}), \text{ for } v \in \hat{C}_0^\infty .$$

Notice that this time, since $\hat{P}_h$ is periodic and C^∞ we have $\exp(t\hat{P}_h) \in M_p$ for $1 \le p \le \infty$ (cf. the beginning of Section 1.4) so that $E_h(t)$ is automatically a bounded operator in L_p for $1 \le p \le \infty$. Developing its symbol in an absolutely convergent Fourier series,

$$\exp(t\hat{P}_h(\xi)) = \sum_\beta e_\beta(h,t)e^{i\langle\xi,h\beta\rangle} ,$$

the solution operator may be represented as

$$E_h(t)v = \sum_\beta e_\beta(h,t)T_h^\beta v ,$$

We say that the semi-discrete problem is well posed in L_p if there exists $h_1 > 0$ and for any $T > 0$ a constant C such that

$$\|E_h(t)v\|_p \le C\|v\|_p, \text{ for } t \le T,\ h \le h_1,\ v \in \hat{C}_0^\infty .$$

In the same way as for the continuous problem (cf. (1.3)), if this holds for some $T > 0$ we have for all $t > 0$ and $h \le h_1$,

$$\|E_h(t)v\|_p \le C_1 e^{C_2 t} \|v\|_p .$$

We have the following analogue of Proposition 1.1.

Proposition 2.2. The semi-discrete problem (2.5) is well posed in L_p if and only if there exists an $h_1 > 0$ and for each $T > 0$ a constant C such that

$$M_p(\exp(t\hat{P}_h)) \le C, \text{ for } t \le T,\ h \le h_1 . \tag{2.6}$$

In particular, (2.5) is well posed in L_2 if and only if for some constant C,

$$\operatorname{Re} \hat{P}_h(\xi) \leq C, \text{ for } \xi \in R^d, \; h \leq h_1 .$$

Proof. The first part is obvious from the definitions. The second follows since

$$M_2(\exp(t\hat{P}_h)) = \exp(t \sup_\xi \operatorname{Re} \hat{P}_h(\xi)) .$$

The well-posedness of the semi-discrete problem thus depends on the choice of the approximation P_h. However, we notice the following:

Proposition 2.3. Let P_h consistent with P. Then a necessary condition for the semi-discrete problem (2.5) to be well posed in L_p is that the continuous problem (1.1) is well posed in L_p.

Proof. From Proposition 2.1 we see that

$$\lim_{h \to 0} \exp(t\hat{P}_h(\xi)) = \exp(t\hat{P}(\xi)), \text{ for } \xi \in R^d,$$

uniformly on compact subsets of R^d. The result hence follows from (2.6) and Theorem 1.2.6.

It is again possible to prove maximum-norm estimates for the solution when the initial value problem (2.5) is well posed only in L_2; we have the following analogue of Theorem 1.1.

Theorem 2.1. Let the initial value problem (2.5) be well posed in L_2. Then for each $T > 0$ there exists a constant C such that for $v \in B_2^{d/2,1}$,

$$\|\mathbb{E}_h(t)v\|_\infty \leq C\|v\|_{B_2^{d/2,1}}, \text{ for } t \leq T, \; h \leq h_1 .$$

Proof. In the same way as Theorem 1.1, this is an immediate consequence of Corollary 2.2.1.

We shall now consider schemes which are discrete also with respect to time. (For simplicity we shall consider only single step methods.) For this purpose, we introduce a time step k tied to h by the relation

(2.7) $$k/h^M = \lambda = \text{constant}.$$

We want to approximate the exact solution $u(nk) = E(nk)v$ of (1.1) at $t = nk$ by u^n, $n = 0,1,2,\ldots$, defined by the discrete-in-time initial value problem

(2.8) $$A_h u^{n+1} = B_h u^n, \text{ for } n = 0,1,\ldots \qquad u^0 = v.$$

Here A_h and B_h are finite difference operators of the form

$$A_h v = \sum_\beta a_\beta(h) T_h^\beta v,$$

$$B_h v = \sum_\beta b_\beta(h) T_h^\beta v,$$

where the $a_\beta(h)$ and $b_\beta(h)$ are polynomials in h and where the summations are finite. If A_h is the identity operator the finite difference scheme corresponding to (2.8) is called explicit, otherwise implicit. Taking Fourier transforms with respect to x, (2.8) reduces to

(2.9) $$\hat{A}_h \hat{u}^{n+1} = \hat{B}_h \hat{u}^n, \text{ for } n = 0,1,2,\ldots \qquad \hat{u}^0 = \hat{v},$$

where

$$\hat{A}_h(\xi) = \sum_\beta a_\beta(h) e^{i\langle\xi,h\beta\rangle},$$

$$\hat{B}_h(\xi) = \sum_\beta b_\beta(h) e^{i\langle\xi,h\beta\rangle}.$$

We shall assume that

$$\sum_\beta a_\beta(0) e^{i\langle\xi,\beta\rangle} \neq 0, \text{ for } \xi \in R^d,$$

which implies that for $h \leq h_0$, say, $\hat{A}_h(\xi)$ is bounded away from zero. Solving for $\hat{u}^{n+1}$ in (2.9) we obtain for $v \in \hat{C}_0^\infty$ that $u^n = E_k^n v$, where

$$E_k v = \mathcal{F}^{-1}(\hat{E}_k \hat{v}) \quad \text{with} \quad \hat{E}_k(\xi) = \hat{B}_h(\xi)/\hat{A}_h(\xi) .$$

We say that E_k is consistent with (1.1) if

$$(2.10) \qquad \hat{E}_k(h^{-1}\xi) = \exp(k\hat{P}(h^{-1}\xi) + o(h^M + |\xi|^M)), \text{ as } h,\xi \to 0.$$

Since its symbol $\hat{E}_k$ is C^∞ and periodic, the operator E_k is bounded in L_p. Developing $\hat{E}_k$ into its absolutely convergent Fourier series,

$$\hat{E}_k(\xi) = \sum_\beta e_\beta(h) e^{i\langle\xi, h\beta\rangle} ,$$

we have the representation

$$E_k v = \sum_\beta e_\beta(h) T_h^\beta v .$$

We say that E_k is stable in L_p if there exists $h_1 > 0$ such that the powers of E_k are bounded in such a way that for any $T > 0$ there exists a constant C such that

$$\|E_k^n v\|_p \leq C\|v\|_p, \text{ for } nk \leq T, \; h \leq h_1, \; v \in \hat{C}_0^\infty .$$

We shall express this property in terms of Fourier multipliers. For later convenience we introduce the 2π-periodic characteristic function

$$e_k(\xi) = \hat{E}_k(h^{-1}\xi),$$

and we then have the following characterization.

<u>Proposition 2.4</u>. The finite difference operator E_k is stable in L_p if and only if there exists a positive h_1 and for any $T > 0$ a constant C such that

$$(2.11) \qquad M_p(\hat{E}_k^n) = M_p(e_k^n) \leq C, \text{ for } nk \leq T, \; h \leq h_1.$$

In particular, E_k is stable in L_2 if and only if there exists a constant C such that

$$\sup_{\xi} |e_k(\xi)| \leq 1+Ck, \text{ for } h \leq h_1. \tag{2.12}$$

Proof. The first part follows at once from the definitions, and the second from the fact that

$$M_2(e_k^n) = (\sup_{\xi} |e_k(\xi)|)^n \ ;$$

the latter quantity is bounded for $nk \leq T$ if and only if (2.12) holds.

Notice that if $e_k(\xi) = e(\xi)$ is independent of k, then the condition (2.12) reduces to $|e(\xi)| \leq 1$. In this case we have

$$\|\mathbb{E}_k^n v\|_2 \leq \|v\|_2, \text{ for } n = 1,2,\ldots \ .$$

The following is an analogue of Proposition 2.3.

Proposition 2.5. Let E_k be consistent with (1.1). Then a necessary condition for E_k to be stable in L_p is that the problem (1.1) is well posed in L_p.

Proof. By (2.10) we have, uniformly on compact subsets of R^d,

$$\lim_{\substack{h \to 0 \\ nk=t}} \hat{E}_k(\xi)^n = \exp(it\hat{P}(\xi)) \ ,$$

and hence the result follows by (2.11) and Theorem 1.2.6.

Corresponding to Theorem 2.1 we have the following:

Theorem 2.2. Assume that E_k is stable in L_2. Then for each $T > 0$ there exists a constant C such that for $v \in B_2^{d/2,1}$,

$$\|\mathbb{E}_k^n v\|_\infty \leq C\|v\|_{B_2^{d/2,1}}, \text{ for } nk \leq T, \ h \leq h_1.$$

Proof. This follows again at once from Corollary 2.2.1.

One way of constructing a consistent finite difference operator E_k is to let P_h be of the form (2.1), to approximate $\exp(z)$ by a rational function $r = b/a$, where a and b are polynomials and $a(z) \neq 0$ for $z \in \{k\hat{P}_h(\xi):\xi \in R^d,\ h \leq h_1\}$, and then to choose $\hat{E}_k = \hat{B}_h/\hat{A}_h$ where $\hat{A}_h = a(k\hat{P}_h)$ and $\hat{B}_h = b(k\hat{P}_h)$. In particular, with $r(z) = 1+z,\ (1-z)^{-1},\ (1+\frac{1}{2}z)(1-\frac{1}{2}z)^{-1}$, we obtain the following simple explicit, implicit and trapezoidal (Crank-Nicholson) schemes, namely

$$u^{n+1} = (I + kP_h)u^n,$$

$$(I - kP_h)u^{n+1} = u^n,$$

$$(I - \tfrac{1}{2}kP_h)u^{n+1} = (I + \tfrac{1}{2}kP_h)u^n,$$

respectively. More accurate approximations may be obtained by using higher order Padé approximations of $\exp(z)$.

Assuming that the initial value problem (1.1) is correctly posed in L_2, we have for P_h defined by symmetric difference quotients, i.e. with $\hat{P}_h(\xi) = \hat{P}(h^{-1}\sin(h\xi))$, that the two latter schemes are stable in L_2 for any choice of λ (unconditionally stable). For the implicit operator,

$$|e_k(\xi)| = |1 - k\hat{P}(h^{-1}\sin\xi)|^{-1} \leq (1 - k\sup_{\xi}\operatorname{Re}\hat{P}(\xi))^{-1} \leq (1 - Ck)^{-1} \leq 1 + C_1k$$

for k small, and for the trapezoidal scheme,

$$|e_k(\xi)|^2 = |(1+\tfrac{1}{2}k\hat{P}(h^{-1}\sin\xi))(1-\tfrac{1}{2}k\hat{P}(h^{-1}\sin\xi))^{-1}|^2$$

$$= 1 + 2k\operatorname{Re}\hat{P}(h^{-1}\sin\xi)|(1-\tfrac{1}{2}k\hat{P}(h^{-1}\sin\xi))^{-1}|^2,$$

so that in both cases (2.12) holds. For the explicit scheme we have

$$|e_k(\xi)|^2 = 1 + 2k\operatorname{Re}\hat{P}(h^{-1}\sin\xi) + k^2|\hat{P}(h^{-1}\sin\xi)|^2,$$

and from this it is easily seen that in general (2.12) is not satisfied. For example, for the heat equation (then $\hat{P}(\xi) = -|\xi|^2$) we then have

$$\sup_{\xi} |e_k(\xi)| = \sup_{\xi} |1-\lambda \sin^2\xi| = \max(|1-\lambda|,1) .$$

In this case (2.12) holds if and only if $\lambda \leq 2$. (The more natural explicit scheme $E_k v = \sum_{j=1}^{d} (T_h^{e_j} - 2I + T_h^{-e_j})v$ corresponding to $e(\xi) = 2 \sum_{j=1}^{d} (1-\cos \xi_j)$ is similarly stable in L_2 if and only if $\lambda \leq 1/2$.)

To avoid unnecessary complications we will assume in the following without explicit mention that h is sufficiently small.

3.3. Accuracy and convergence.

In this section we shall introduce the concept of the order of convergence of a finite difference operator and obtain some related simple convergence estimates for problems which are well posed in L_2.

Consider first the initial value problem (1.1) with solution operator $E(t)$, and a semi-discrete difference approximation $E_h(t)$ where P_h is consistent with P.

We say that P_h is accurate of order $\mu > 0$ if

$$(3.1) \qquad h^M \hat{P}_h(h^{-1}\xi) = h^M \hat{P}(h^{-1}\xi) + o(h^{M+\mu} + |\xi|^{M+\mu}), \text{ as } h,\xi \to 0.$$

Our first convergence result concerns smooth initial data.

Theorem 3.1. Assume that (1.1) and (2.5) are well posed in L_2 and that P_h is accurate of order μ. Then for each $T > 0$ there exists a constant C such that for $v \in W_2^{M+\mu}$,

$$\|E_h(t)v - E(t)v\|_2 \leq Ch^\mu \|v\|_{W_2^{M+\mu}} , \text{ for } t \leq T.$$

<u>Proof</u>. We may write

$$E_h(t)v - E(t)v = \mathcal{F}^{-1}((\exp(t\hat{P}_h(\xi)) - \exp(t\hat{P}(\xi)))\hat{v}).$$

By (3.1) and Propositions 1.1 and 2.2 we have for $h|\xi| \leq \varepsilon$, say

$$|\exp(t\hat{P}_h(\xi)) - \exp(t\hat{P}(\xi))| \leq C|\hat{P}_h(\xi) - \hat{P}(\xi)| \leq Ch^{\mu}(1 + |\xi|^{M+\mu}).$$

On the other hand, for $h|\xi| > \varepsilon$ we have by Propositions 1.1 and 2.2

$$|\exp(t\hat{P}_h(\xi)) - \exp(t\hat{P}(\xi))| \leq C \leq C(h|\xi|)^{M+\mu} \leq Ch^{\mu}(1 + |\xi|^{M+\mu}).$$

Hence by Parseval's formula we obtain

$$\|E_h(t)v - E(t)v\|_2^2 \leq Ch^{2\mu} \int (1 + |\xi|^{M+\mu})^2 |\hat{v}(\xi)|^2 d\xi \leq Ch^{2\mu} \|v\|^2_{W_2^{M+\mu}},$$

which proves the theorem.

For less stringent regularity assumptions on the initial data we may obtain correspondingly weaker convergence estimates.

<u>Theorem 3.2</u>. Under the assumptions of Theorem 3.1, let $0 < s < M+\mu$. Then for each $T > 0$ there exists a constant C such that for $v \in B_2^{s,\infty}$,

$$\|E_h(t)v - E(t)v\|_2 \leq Ch^{\frac{s\mu}{\mu+M}} \|v\|_{B_2^{s,\infty}}, \quad \text{for } t \leq T.$$

<u>Proof</u>. Notice that by the well-posedness in L_2 of (1.1) and (2.5),

$$\|E_h(t)v - E(t)v\|_2 \leq C\|v\|_2, \quad \text{for } t \leq T.$$

The result hence follows from Theorem 3.1 by interpolation (Corollary 2.5.1).

We can now prove the following analogue of Theorem 3.1 in the maximum norm.

<u>Theorem 3.3</u>. Under the assumption of Theorem 3.1, there exists for each $T > 0$ a

constant C such that for $v \in B_2^{d/2+M+\mu,1}$

$$\|E_h(t)v - E(t)v\|_\infty \leq Ch^\mu \|v\|_{B_2^{d/2+M+\mu,1}}, \quad \text{for} \quad t \leq T.$$

Proof. Since

$$(E_h(t) - E(t))v = \mathcal{F}^{-1}((\exp(t\hat{P}_h) - \exp(t\hat{P}))\hat{v}),$$

and since by the proof of Theorem 3.1 we have for $t \leq T$,

$$|\exp(t\hat{P}_h(\xi)) - \exp(t\hat{P}(\xi))| \leq Ch^\mu(1 + |\xi|^{M+\mu}),$$

the result follows at once by Corollary 2.2.1.

For less smooth initial data we have correspondingly:

Theorem 3.4. Under the assumptions of Theorem 3.1, let $0 < s < M+\mu$. Then for each $T > 0$ there exists a constant C such that for $v \in B_2^{d/2+s,\infty}$,

$$\|E_h(t)v - E(t)v\|_\infty \leq Ch^{\frac{s\mu}{\mu+M}} \|v\|_{B_2^{d/2+s,\infty}}, \quad \text{for} \quad t \leq T.$$

Proof. By Theorems 1.1 and 2.1 we have

$$\|E_h(t)v - E(t)v\|_\infty \leq C\|v\|_{B_2^{d/2,1}}.$$

Interpolation between this and the result of Theorem 3.3 proves the desired result.

We next consider the analogous results for completely discrete finite difference operators. We say that the finite difference operator E_k is accurate of order $\mu > 0$ if

$$(3.2) \qquad \hat{E}_k(h^{-1}\xi) = \exp(k\hat{P}(h^{-1}\xi) + O(h^{M+\mu} + |\xi|^{M+\mu})), \text{ as } h,\xi \to 0.$$

In particular, E_k is then consistent with (1.1).

Theorem 3.5. Assume that (1.1) is well posed in L_2, that E_k is stable in L_2, and that E_k is accurate of order μ. Then for each $T > 0$ there exists a constant C such that for $v \in W_2^{M+\mu}$,

$$\|E_k^n v - E(nk)v\|_2 \leq Ch^\mu \|v\|_{W_2^{M+\mu}}, \text{ for } nk \leq T.$$

Proof. In the same way as in the proof of Theorem 3.1 we have by (3.2) and Propositions 1.1 and 2.4 that

$$|\hat{E}_k(\xi)^n - \exp(nk\hat{P}(\xi))| \leq Ch^\mu(1 + |\xi|^{M+\mu}), \text{ for } \xi \in R^d,$$

and the result again follows by Parseval's relation.

Similarly to Theorems 3.2, 3.3, and 3.4 in the semi-discrete situation we now easily obtain the following sequence of analogous results.

Theorem 3.6. Under the assumptions of Theorem 3.5, let $0 < s < M+\mu$. Then for each $T > 0$ there exists a constant C such that for $v \in B_2^{s,\infty}$,

$$\|E_k^n v - E(nk)v\|_2 \leq Ch^{\frac{s\mu}{\mu+M}} \|v\|_{B_2^{s,\infty}}, \text{ for } nk \leq T.$$

Theorem 3.7. Under the assumptions of Theorem 3.5, for each $T > 0$ there exists a constant C such that for $v \in B_2^{d/2+M+\mu,1}$,

$$\|E_k^n v - E(nk)v\|_\infty \leq Ch^\mu \|v\|_{B_2^{d/2+M+\mu,1}}, \text{ for } nk \leq T.$$

Theorem 3.8. Under the assumptions of Theorem 3.5, let $0 < s < M+\mu$. Then for each $T > 0$ there exists a constant C such that for $v \in B_2^{d/2+s,\infty}$

$$\|E_k^n v - E(nk)v\|_\infty \leq Ch^{\frac{s\mu}{\mu+M}} \|v\|_{B_2^{d/2+s,\infty}}, \text{ for } nk \leq T.$$

References.

For basic material on initial value problems and associated finite difference schemes, including discussions of stability and convergence properties, see [1], [2], [3] and references.

1. R.D. Richtmyer and K.W. Morton, Difference Methods for Initial Value Problems, Interscience, New York 1967.

2. J. Peetre and V. Thomée, On the rate of convergence for discrete initial value problems, Math. Scand. 21 (1967), 159-176.

3. V. Thomée, Stability theory for partial difference operators, SIAM Rev. 11 (1969), 152-195.

CHAPTER 4. THE HEAT EQUATION.

In this chapter we shall consider semi-discrete approximations to the initial value problem for the heat equation. In particular, we shall investigate how the rate of convergence of the approximate solution to the exact solution depends on the smoothness of the initial function. The results are expressed in terms of Besov spaces and the proofs use the techniques developed in Chapters 1 and 2. Since our methods do not depend strongly on the number of space variables we consider (for simplicity) only the one-dimensional problem.

In Section 1 we prove convergence estimates in L_p with the smoothness of the data measured in the same L_p space. We also estimate the rate of convergence of difference quotients of the approximate solution to derivatives of the solution of the continuous problem. In Section 2 we derive various inverse results. On the one hand these show that the convergence estimates of Section 1 are in a certain sense best possible, and on the other hand they motivate another type of convergence estimates, presented in Section 3, in which the error is measured in the maximum norm but the smoothness of the data in L_1. Finally, in Section 4 we consider the effect of a preliminary smoothing of the initial data.

4.1. Convergence estimates in L_p.

We shall consider the initial value problem for the one-dimensional heat equation

$$\frac{\partial u}{\partial t} = \frac{\partial^2 u}{\partial x^2}, \quad \text{for } x \in R,\ t > 0, \tag{1.1}$$
$$u(x,0) = v(x) .$$

Recall from Section 3.1 that the solution operator of this problem is defined by

$$E(t)v = \mathcal{F}^{-1}(\exp(-t\xi^2)\hat{v}),$$

and that the problem is well posed in L_p for $1 \le p \le \infty$. In the present case the solution is smooth for positive t even when the initial function is not, as the following result shows:

Theorem 1.1. Let $1 \le p \le \infty$, and $\alpha \ge 0$ be an integer. Then with $E(t)$ as above there exists a constant C such that for $v \in L_p$,

$$\|D^\alpha E(t)v\|_p \le Ct^{-\alpha/2}\|v\|_p, \text{ for } t > 0.$$

Proof. We may write

$$D^\alpha E(t)v = \mathcal{F}^{-1}((i\xi)^\alpha \exp(-t\xi^2)\hat{v}).$$

By Theorem 1.2.8 we have for $t > 0$,

$$M_p((i\xi)^\alpha \exp(-t\xi^2)) = t^{-\alpha/2} M_p(\xi^\alpha \exp(-\xi^2)),$$

and since $\xi^\alpha \exp(-\xi^2) \in S \subseteq M_p$ for $1 \le p \le \infty$, the result follows.

Consider now a finite difference approximation $P_h = h^{-2} \sum_\beta p_\beta T_h^\beta$ of (note the minus sign introduced for convenience in this chapter) $-D^2 = -(d/dx)^2$, where for simplicity the coefficients are assumed independent of h. The solution of the corresponding semi-discrete problem (3.2.5) can then be written

$$E_h(t)v = \mathcal{F}^{-1}(\exp(-t\hat{P}_h)\hat{v}) = \mathcal{F}^{-1}(\exp(-th^{-2}p(h\xi))\hat{v}),$$

where we have introduced the trigonometric polynomial

$$p(\xi) = h^2 \hat{P}_h(h^{-1}\xi) = \sum_\beta p_\beta e^{i\beta\xi} .$$

(In fact, all results below are valid for $p(\xi)$ 2π-periodic and real analytic.)

In the following we shall assume that P_h is accurate of order μ. This condition reduces in the present case to

(1.2) $$p(\xi) = \xi^2 + O(|\xi|^{2+\mu}), \text{ as } \xi \to 0.$$

We shall also assume that the semi-discrete problem is parabolic, which we define to mean that there exists a positive constant c such that

(1.3) $$\operatorname{Re} p(\xi) \geq c\xi^2, \text{ for } |\xi| \leq \pi.$$

Notice that for small values of $|\xi|$ such an inequality follows from (1.2).

For α a positive integer, let $Q_h^\alpha = h^{-\alpha} \sum_\beta q_\beta T_h^\beta$ be a difference operator with coefficients independent of h. Assume that Q_h^α approximates $D^\alpha = (\frac{d}{dx})^\alpha$ with order of accuracy μ, that is, if $q_\alpha(\xi) = \sum_\beta q_\beta e^{i\beta\xi}$, let

(1.4) $$q_\alpha(\xi) = (i\xi)^\alpha + O(|\xi|^{\alpha+\mu}), \text{ as } \xi \to 0.$$

To simplify the presentation below, Q_h^0 will denote the identity operator, and $q_0(\xi) = 1$. Notice that

$$Q_h^\alpha v = \mathcal{F}^{-1}(h^{-\alpha} q_\alpha(h\xi)\hat{v}) .$$

The following is now a discrete analogue of Theorem 1.1.

<u>Theorem 1.2</u>. Let $1 \leq p \leq \infty$, and $\alpha \geq 0$ be an integer. Let $E_h(t)$ and Q_h^α be as above. Then there exists a constant C such that for $v \in L_p$,

$$\|Q_h^\alpha E_h(t)v\|_p \leq Ct^{-\alpha/2} \|v\|_p, \text{ for } t > 0.$$

<u>Proof</u>. Since

$$Q_h^\alpha E_h(t)v = \mathcal{F}^{-1}(h^{-\alpha} q_\alpha(h\xi)\exp(-th^{-2}p(h\xi))\hat{v}) ,$$

it is sufficient to prove that

$$M_p(h^{-\alpha} q_\alpha(h\xi)\exp(-th^{-2}p(h\xi))) \leq Ct^{-\alpha/2} ,$$

or by Theorem 1.2.8 and a change of variable $h\xi \to \xi$, that

$$M_p(q_\alpha \exp(-th^{-2}p)) \leq C(th^{-2})^{-\alpha/2}.$$

With η as in Theorem 1.4.1 and $\tau = th^{-2}$ it therefore suffices to show that for $\tau > 0$,

$$(1.5) \qquad M_p(d_{\alpha\tau}) \leq C\tau^{-\alpha/2} \quad \text{where} \quad d_{\alpha\tau} = \eta q_\alpha \exp(-\tau p),$$

and since $M_p \supseteq M_\infty$, it is enough to prove (1.5) for $p = \infty$. This will now be done by using the Carlson-Beurling inequality (Theorem 1.3.1).

Let first α be positive. By periodicity we may assume that (1.3) holds (possibly with a smaller c) on $\mathrm{supp}(\eta)$. Since by (1.4),

$$|q_\alpha(\xi)| \leq C|\xi|^\alpha \quad \text{for} \quad \xi \in \mathrm{supp}(\eta),$$

we have by (1.3),

$$(1.6) \qquad \|d_{\alpha\tau}\|_2 \leq C\left(\int \xi^{2\alpha} \exp(-2c\tau\xi^2)d\xi\right)^{1/2} = C\tau^{-\frac{\alpha}{2}-\frac{1}{4}}.$$

It also follows by (1.4) that for $\xi \in \mathrm{supp}(\eta)$,

$$(1.7) \qquad \left|\frac{d}{d\xi} q_\alpha(\xi)\right| \leq C|\xi|^{\alpha-1},$$

and by (1.2) and (1.3) that for these ξ,

$$(1.8) \qquad \left|\frac{d}{d\xi} \exp(-\tau p(\xi))\right| \leq C\tau|\xi| \exp(-c\tau\xi^2),$$

We conclude that

$$\left|\frac{d}{d\xi} d_{\alpha\tau}(\xi)\right| \leq C(|\xi|^{\alpha-1} + \tau|\xi|^{\alpha+1})\exp(-c\tau\xi^2),$$

and hence by integration,

$$(1.9) \qquad \left\|\frac{d}{d\xi} d_{\alpha\tau}\right\|_2 \leq C\tau^{-\frac{\alpha}{2}+\frac{1}{4}}.$$

By the Carlson-Beurling inequality, (1.6) and (1.9) prove (1.5) for $p = \infty$.

Consider now the case $\alpha = 0$. Since $q_0 = 1$ we obtain

$$\|d_{0\tau}\|_2 \leq C\Big(\int_{\mathrm{supp}(\eta)} \exp(-2c\tau\xi^2)d\xi\Big)^{1/2} \leq C\min(1,\tau^{-1/4}),$$

and

$$\Big\|\frac{d}{d\xi}\, d_{0\tau}\Big\|_2 \leq C\Big(\int_{\mathrm{supp}(\eta)} (1+\tau|\xi|)^2\exp(-2c\tau\xi^2)d\xi\Big)^{1/2} \leq C(1+\tau^{1/4})$$

from which (1.5) again follows for $p = \infty$ by the Carlson-Beurling inequality.

This completes the proof of the theorem.

We next prove estimates for the rate of convergence of $Q_h^{\alpha}E_h(t)v$ to $D^{\alpha}E(t)v$.

<u>Theorem 1.3</u>. Let $1 \leq p \leq \infty$, and $\alpha \geq 0$ be an integer. Let $E(t)$, $E_h(t)$ and Q_h^{α} be as above. In particular, let P_h and Q_h^{α} be accurate of order μ (and let Q_h^0 denote the identity operator). Then there exists a constant C such that for $v \in B_p^{\mu,\infty}$,

$$\|Q_h^{\alpha}E_h(t)v - D^{\alpha}E(t)v\|_p \leq Ct^{-\alpha/2}h^{\mu}\|v\|_{B_p^{\mu,\infty}}, \quad \text{for } t > 0.$$

<u>Proof</u>. With

$$g_{\tau}(\xi) = q_{\alpha}(\xi)\exp(-\tau p(\xi)) - (i\xi)^{\alpha}\exp(-\tau\xi^2),$$

we have

$$Q_h^{\alpha}E_h(t)v - D^{\alpha}E_h(t)v = \mathcal{F}^{-1}(h^{-\alpha}g_{th^{-2}}(h\xi)\hat{v}).$$

By Lemma 2.6.2 it is therefore sufficient to prove that for $\tau > 0$,

$$\dot{M}_p^{\mu,1}(g_{\tau}) = \sum_{j=-\infty}^{\infty} 2^{-\mu j} M_p(\phi_j(h^{-1}\cdot)g_{\tau}) \leq C\tau^{-\alpha/2}h^{\mu}. \tag{1.10}$$

Again it is enough to consider $p = \infty$, and we shall then want to estimate

$$m_j = M_{\infty}(\phi_j(h^{-1}\cdot)g_{\tau}).$$

The estimates for the m_j:s will be divided into two cases, depending on the size of $2^j h$. We first consider $2^j h \leq \pi/2$, and write $g_\tau = g_{\tau,1} + g_{\tau,2}$ with

$$g_{\tau,1}(\xi) = (i\xi)^\alpha(\exp(-\tau p(\xi)) - \exp(-\tau\xi^2))$$

$$= \tau \int_0^1 (i\xi)^\alpha(\xi^2 - p(\xi))\exp(-(1-s)\tau\xi^2 - s\tau p(\xi))ds ,$$

$$g_{\tau,2}(\xi) = (q_\alpha(\xi) - (i\xi)^\alpha)\exp(-\tau p(\xi)).$$

Setting $b_\alpha(\xi) = (i\xi)^\alpha(\xi^2 - p(\xi))$ and $e_{s,\tau}(\xi) = \exp(-(1-s)\tau\xi^2 - s\tau p(\xi))$, we obtain by the triangle inequality in M_∞,

$$(1.11) \qquad m_j^{(1)} = M_\infty(\phi_j(h^{-1}\cdot)g_{\tau,1}) \leq \tau \int_0^1 M_\infty(\phi_j(h^{-1}\cdot)b_\alpha e_{s,\tau})ds .$$

Let now $\omega = 2^j h$, so that $\phi_j(h^{-1}\xi) = \phi(\omega^{-1}\xi)$. Then by (1.2) and (1.3) we have

$$(1.12) \qquad |b_\alpha(\xi)e_{s,\tau}(\xi)| \leq C|\xi|^{\alpha+\mu+2}\exp(-c\tau|\xi|^2)$$

$$\leq C\omega^{\alpha+\mu+2}\exp(-c\tau\omega^2), \text{ for } \tfrac{1}{2}\omega \leq |\xi| \leq 2\omega ,$$

and hence, since $\operatorname{supp}(\phi(\omega^{-1}\cdot)) \subseteq \{\xi: \frac{1}{2}\omega < |\xi| < 2\omega\}$,

$$\|\phi(\omega^{-1}\cdot)b_\alpha e_{s,\tau}\|_2 \leq C\omega^{\alpha+\mu+5/2}\exp(-c\tau\omega^2).$$

By (1.4), (1.6), (1.8) and (1.12), we also have

$$\left|\frac{d}{d\xi}(\phi(\omega^{-1}\xi)b_\alpha(\xi)e_{s,\tau}(\xi))\right| \leq C(\omega^{-1}|\xi|^{\alpha+\mu+2} + \tau|\xi|^{\alpha+\mu+3} + |\xi|^{\alpha+\mu+1})\exp(-c\tau\xi^2)$$

$$\leq C\omega^{\alpha+\mu+1}\exp(-c\,\tau\omega^2).$$

It follows that

$$\left\|\frac{d}{d\xi}[\phi(\omega^{-1}\cdot)b_\alpha e_{s,\tau}]\right\|_2 \leq C\omega^{\alpha+\mu+3/2}\exp(-c\tau\omega^2).$$

The Carlson-Beurling inequality and (1.11) then prove that for $\omega = 2^j h \leq \pi/2$,

$$m_j^{(1)} \leq C\tau\omega^{\alpha+\mu+2}\exp(-c\tau\omega^2) \leq C\omega^{\alpha+\mu}\exp(-c\tau\omega^2).$$

Since for $j = 0,1$,

$$\left(\frac{d}{d\xi}\right)^j (q_\alpha(\xi) - (i\xi)^\alpha) = O(|\xi|^{\alpha+\mu-j}), \text{ as } \xi \to 0,$$

we have for $\alpha > 0$ the estimates

$$|g_{\tau,2}(\xi)| \leq C|\xi|^{\alpha+\mu}\exp(-c\tau|\xi|^2),$$

and

$$\left|\frac{d}{d\xi} g_{\tau,2}(\xi)\right| \leq C(|\xi|^{\alpha+\mu-1} + \tau|\xi|^{\alpha+\mu+1})\exp(-c\tau|\xi|^2).$$

As above, the Carlson-Beurling inequality gives with $\omega = 2^j h \leq \pi/2$,

$$m_j^{(2)} = M_\infty(\phi(\omega^{-1}\cdot)g_{\tau,2}) \leq C\omega^{\alpha+\mu}\exp(-c\tau\omega^2).$$

We also notice that since $q_0(\xi) = 1$, $g_{\tau,2} = 0$ for $\alpha = 0$. Together with the above estimates, this proves that for $\omega = 2^j h \leq \pi/2$,

$$(1.13) \qquad m_j \leq \begin{cases} C\omega^{\alpha+\mu}\exp(-c\tau\omega^2), & \text{for } \alpha > 0, \\ C\tau\omega^{\mu+2}\exp(-c\tau\omega^2), & \text{for } \alpha = 0. \end{cases}$$

In the case $2^j h \geq \pi/2$, we have by Theorem 1.4.1, (1.5) and since $\xi^\alpha \exp(-\xi^2) \in S \subset M_\infty$,

$$(1.14) \qquad m_j \leq M_\infty(q_\alpha \exp(-\tau p)) + M_\infty(\xi^\alpha \exp(-\tau\xi^2)) \leq C\tau^{-\alpha/2}.$$

Let now j_0 be the largest integer such that $2^{j_0}h \leq \pi/2$, Then (1.13) and (1.14) yield for $\alpha > 0$,

$$\dot{M}_p^{\mu,1}(g_\tau) \leq \sum_{-\infty}^{\infty} 2^{-\mu j} m_j \leq C\{h^\mu \sum_{-\infty}^{j_0} (2^j h)^\alpha \exp(-c\tau(2^j h)^2) + \tau^{-\alpha/2} \sum_{j_0+1}^{\infty} 2^{-\mu j}\}$$

$$\leq Ch^\mu \tau^{-\alpha/2}\left(\int_0^\infty \xi^{\alpha-1}\exp(-c\xi^2)d\xi + 1\right) \leq Ch^\mu\tau^{-\alpha/2},$$

and for $\alpha = 0$,

$$\dot{M}_p^{\mu,1}(g_\tau) \le C\{h^\mu \sum_{-\infty}^{j_0} \tau(2^j h)^2 \exp(-c\tau(2^j h)^2) + \sum_{j_0+1}^{\infty} 2^{-\mu j}\}$$

$$\le Ch^\mu\{\int_0^\infty \xi \exp(-c\xi^2)d\xi + 1\} \le Ch^\mu .$$

This proves (1.10) and hence completes the proof of the theorem.

The proof above shows that the norm $\|\cdot\|_{B_p^{\mu,\infty}}$ in the convergence estimate can be replaced by the seminorm $\|\cdot\|_{\dot{B}_p^{\mu,\infty}}$. This depends on the fact that the operator P is homogeneous and the coefficients of P_h are independent of h. Similar remarks will apply frequently throughout these notes.

Notice that for $\alpha = 0$ and $p = 2$, the regularity requirements on the initial data in Theorem 1.3 in order to obtain convergence of order h^μ are weaker than in the more general result of Theorem 3.3.1.

By interpolation we now obtain the following result for less regular initial data:

<u>Theorem 1.4</u>. Under the assumptions of Theorem 1.3, let $0 < s \le \mu$. Then there exists a constant C such that for $v \in B_p^{s,\infty}$,

$$\|Q_h^\alpha E_h(t)v - D^\alpha E(t)v\|_p \le Ct^{-\alpha/2} h^s \|v\|_{B_p^{s,\infty}}, \text{ for } t > 0.$$

<u>Proof</u>. By Theorems 1.1 and 1.2,

$$\|Q_h^\alpha E_h(t)v - D^\alpha E(t)v\|_p \le \|Q_h^\alpha E_h(t)v\|_p + \|D^\alpha E(t)v\|_p \le C\tau^{-\alpha/2}\|v\|_p ,$$

and interpolation (Corollary 2.5.1) between this and the result of Theorem 1.3 proves the theorem.

4.2. Inverse results.

In this section we shall present inverse results to the estimates of Section 1, that is, results which state that if a certain rate of convergence holds for a particular initial function, then this function must have a certain degree of smoothness. For simplicity, we shall only treat the case when convergence is measured in the maximum-norm.

Throughout this section, let $E(t)$ be the solution operator of (1.1), and $E_h(t)$ the solution operator of a corresponding semi-discrete parabolic difference problem, so that $E_h(t)v = \mathcal{F}^{-1}(\exp(-t\hat{P}_h)\hat{v})$, where P_h is consistent with $-D^2 = -(d/dx)^2$. In the proofs below, $F_h(t) = E_h(t) - E(t)$ denotes the error operator.

We begin by proving that Theorem 1.4, for $\alpha = 0$, is in a certain sense best possible.

Theorem 2.1. Let P_h be accurate of order μ, let $0 < s \leq \mu$, and let $t > 0$ be fixed. Then there exists a function $v \in B_\infty^{s,\infty}$ such that

$$\limsup_{h \to 0} h^{-s} \| E_h(t)v - E(t)v \|_\infty > 0.$$

Proof. Without loss of generality we may assume that $t = 1$. Let $G_s(x)$ denote the function defined in Example I of Section 2.4 that is with $\hat{G} \in C_0^\infty(0,1)$,

$$(2.1) \qquad G_s(x) = \Big(\sum_{j=1}^{\infty} e^{ix2^j} 2^{-sj} \Big) G(x) .$$

As was proved in Proposition 2.4.1, $G_s \in B_p^{s,\infty}$, for $1 \leq p \leq \infty$. Let now with F_h the error operator,

$$f_{p,h} = \| F_h(1)\, G_s \|_p .$$

By Hölder's inequality and Theorem 1.4 with $p = 1$, we then have

$$f_{2,h}^2 \leq f_{1,h}\, f_{\infty,h} \leq Ch^s f_{\infty,h} .$$

Hence the conclusion of the theorem will follow if we can prove that

$$\limsup_{h \to 0} h^{-s} f_{2,h} > 0. \tag{2.2}$$

Let $I_j = (2^j, 2^j + 1)$ for $j \geq 1$. By Parseval's relation (2.1), and the periodicity of $\exp(-\hat{P}_h)$, we get

$$2\pi f_{2,h}^2 = \int |\exp(-\hat{P}_h(\xi)) - \exp(-\xi^2)|^2 |\hat{G}_s(\xi)|^2 d\xi$$

$$\geq \int_{I_j} |\exp(-\hat{P}_h(\xi - 2\pi h^{-1})) - \exp(-\xi^2)|^2 \, 2^{-2sj} |\hat{G}(\xi - 2^j)|^2 d\xi . \tag{2.3}$$

Set now $h_j = 2\pi 2^{-j}$. By consistency, we have for large values of j,

$$|\exp(-\hat{P}_{h_j}(\xi - 2\pi h_j^{-1}))| \geq \exp(-\tfrac{1}{2}) > 0, \text{ for } \xi \in I_j.$$

Since clearly $\sup_{I_j} |\exp(-\xi^2)|$ tends to zero as j tends to infinity, we therefore obtain from (2.3) that for j_0 large enough, there is a constant $c > 0$ such that

$$f_{2,h_j}^2 \geq c h_j^{2s} \int_{I_j} |\hat{G}(\xi - 2^j)|^2 d\xi = c h_j^{2s} \|\hat{G}\|_2^2, \text{ for } j \geq j_0.$$

This proves (2.2) and the proof of the theorem is complete.

From now on we shall assume that P_h is accurate of order exactly μ, that is P_h is accurate of order μ and there exists a positive constant c such that for ξ small enough,

$$|p(\xi) - \xi^2| \geq c|\xi|^{\mu+2}, \text{ where } p(\xi) = h^2 \hat{P}_h(h^{-1}\xi) . \tag{2.4}$$

The following theorem then shows that under this assumption we cannot in general expect a better convergence rate than $O(h^{\mu})$.

Theorem 2.2. Assume that P_h is accurate of order exactly μ and let $t > 0$ be fixed. If $v \in C_0^\infty$ and if

(2.5) $$\|E_h(t)v - E(t)v\|_\infty = h^\mu o(1), \text{ as } h \to 0,$$

then v vanishes identically.

Proof. Again, without loss of generality we may assume that $t = 1$. Theorem 1.3 with $p = 1$ and Hölder's inequality give together with (2.5),

(2.6) $$\|F_h(1)v\|_2 \leq \|F_h(1)v\|_1^{1/2}\|F_h(1)v\|_\infty^{1/2} = h^\mu o(1), \text{ as } h \to 0.$$

It is sufficient to prove that

(2.7) $$v_j = \mathcal{F}^{-1}(\phi_j\hat{v}) = 0, \text{ for all } j \in Z.$$

For then $\hat{v}(\xi) = 0$ for $\xi \neq 0$ and since $\hat{v} \in S$ we may conclude that $\hat{v} \equiv 0$. We have

$$\mathcal{F}(F_h(1)v)(\xi) = f_{h^{-2}}(h\xi)\hat{v}(\xi),$$

where now

(2.8) $$f_\tau(\xi) = \exp(-\tau p(\xi)) - \exp(-\tau\xi^2).$$

We may hence write

(2.9) $$f_{h^{-2}}(h\xi) = \exp(-\xi^2)\{\exp[-h^{-2}(p(h\xi) - (h\xi)^2)] - 1\},$$

so that (2.4) shows that for ε sufficiently small, and $2^j h \leq \varepsilon$,

$$|f_{h^{-2}}(h\xi)^{-1}| \leq C \exp(\xi^2)h^{-\mu}|\xi|^{-(2+\mu)} \leq c_j h^{-\mu}, \text{ for } \xi \in \operatorname{supp}(\phi_j).$$

Since

$$v_j = \mathcal{F}^{-1}(\phi_j f_{h^{-2}}(h\cdot)^{-1}\mathcal{F}(F_h(1)v)),$$

we obtain from (2.9) that

$$\|v_j\|_2 \le M_2(\phi_j f_{h^{-2}}(h\cdot)^{-1})\|F_h(1)v\|_2 \le c_j h^{-\mu}\|F_h(1)v\|_2 .$$

Letting h tend to zero, we see that (2.6) implies (2.7), and the theorem is proved.

We shall now present two results in which conclusions about the degree of smoothness of the initial data can be drawn from assumptions on the rate of convergence.

<u>Theorem 2.3</u>. Let $s > 0$, and assume that P_h has order or accuracy exactly μ. If $v \in L_\infty$ and if there is a constant C such that

$$\|E_h(t)v - E(t)v\|_\infty \le Ch^s, \text{ for } h \le 1,\ t \le 1,$$

then $v \in B_\infty^{s,\infty}$.

<u>Proof</u>. Let $v_j = \mathcal{F}^{-1}(\psi_j \hat{v})$, $j \ge 0$. For $j > 0$ and any h we have with f_τ defined by (2.8),

$$v_j = \mathcal{F}^{-1}(\psi_j f_{th^{-2}}(h\xi)^{-1}\mathcal{F}(F_h(t)v)), \tag{2.10}$$

provided $f_{th^{-2}}(h\xi) \neq 0$ on $\operatorname{supp}(\psi_j)$. For ε a given positive number we choose $t_j = h_j^2 = 2^{-2j}\varepsilon^2$. Then

$$f_{t_j h_j^{-2}}(h_j\xi) = f_1(\varepsilon 2^{-j}\xi) = \exp(-p(\varepsilon 2^{-j}\xi)) - \exp(-(\varepsilon 2^{-j}\xi)^2),$$

and by (2.4) this function is non-zero on $\operatorname{supp}(\psi_j)$ for ε small enough, so that

$$M_\infty(\psi_j f_1(\varepsilon 2^{-j}\cdot)^{-1}) = M_\infty(\phi(\varepsilon^{-1}\cdot)f_1^{-1}) < \infty .$$

By (2.10) we conclude that

$$\|v_j\|_\infty \le C\|F_{h_j}(t_j)v\|_\infty \le Ch_j^s \le C2^{-js}. \tag{2.11}$$

If $j = 0$, we have

(2.12) $\quad \|v_0\|_\infty \le M_\infty(\psi_0)\|v\|_\infty < +\infty .$

Together the inequalities (2.11) and (2.12) prove that $v \in B_\infty^{s,\infty}$, which is the desired result.

The result of Theorem 2.3 depends heavily on the assumption that the convergence estimate holds uniformly for small t. In the next theorem the convergence rate is assumed only for a fixed positive t.

Theorem 2.4. Let P_h be consistent with $-D^2$, let $t > 0$ be fixed, and assume that $s > 1$. If $v \in L_\infty$ and if for h small,

$$\|E_h(t)v - E(t)v\|_\infty \le Ch^s,$$

then $v \in B_\infty^{s-1,\infty}$.

Proof. Let $v_j = F^{-1}(\psi_j v)$, $j \ge 0$. Again we may assume that $t = 1$. Let $g \ge 0$ be a function in $C_0^\infty(-1,1)$, such that with $g_m(\xi) = g(\xi-m)$ for $m \in Z$,

$$\sum_{m=-\infty}^{\infty} g_m(\xi) = 1.$$

For $j > 0$ we then have

(2.13) $$\|v_j\|_\infty \le \sum_{2^{j-1} \le |m| \le 2^{j+1}} \|\mathcal{F}^{-1}(\psi_j g_m \hat{v})\|_\infty .$$

In estimating the terms in (2.13) we shall only consider j and m such that $2^{j-1} \le m \le 2^{j+1}$; negative values of m can be treated similarly. We write $h_m = 2\pi/m$, and

$$e_m(\xi) = f_{h_m^{-2}}(h_m\xi) = \exp(-h_m^{-2}p(h_m\xi - 2\pi)) - \exp(-\xi^2) ,$$

where the last equality follows from the periodicity of p.

Notice that

$$(2.14) \qquad -h_m \leq h_m\xi - 2\pi \leq h_m, \text{ for } \xi \in \operatorname{supp}(g_m).$$

Hence by consistency we obtain for large j,

$$\exp(-h_m^2 p(h_m\xi - 2\pi)) \geq c > 0,$$

and since $\exp(-\xi^2)$ tends to zero as ξ tends to infinity, we conclude that for large j,

$$|e_m(\xi)| \geq c > 0, \text{ for } \xi \in \operatorname{supp}(g_m),$$

so that

$$(2.15) \qquad \|g_m e_m^{-1}\|_2 \leq C.$$

Consistency also implies that

$$\left|\frac{d}{d\xi}\, p(\xi)\right| \leq C|\xi|, \text{ as } \xi \to 0,$$

and hence for large j, again using (2.14), that for $\xi \in \operatorname{supp}(g_m)$,

$$\left|\frac{d}{d\xi}\,(e_m(\xi)^{-1})\right| \leq |e_m(\xi)|^{-2}\left(h_m^{-1}\left|\frac{dp}{d\xi}(h_m\xi - 2\pi)\right| + 2|\xi|\, \exp(-\xi^2)\right) \leq C.$$

Thus

$$(2.16) \qquad \left\|\frac{d}{d\xi}(g_m e_m^{-1})\right\|_2 \leq C,$$

and by the Carlson-Beurling inequality, (2.15) and (2.16) give for j large,

$$M_\infty(g_m e_m^{-1}) \leq C.$$

Hence for large j we obtain, using now the assumption of the theorem,

$$\|\mathcal{F}^{-1}(\psi_j g_m \hat{v})\|_\infty = \|\mathcal{F}^{-1}(\psi_j g_m e_m^{-1} \mathcal{F}(F_{h_m}(1)v))\|_\infty \leq M_\infty(g_m e_m^{-1})\|F_{h_m}(1)v\|_\infty$$

$$\leq Ch_m^s \leq C2^{-js}.$$

Since the summation in (2.13) involves less than 2^{j+2} terms, we obtain hence for some j_0,

$$(2.17) \qquad \|v_j\|_\infty \le C2^{-j(s-1)}, \text{ for } j \ge j_0.$$

On the other hand, we have for $j < j_0$,

$$(2.18) \qquad \|v_j\|_\infty \le \max[M_\infty(\psi_0), M_\infty(\phi)]\,\|v\|_\infty \le C \le C2^{-j(s-1)}.$$

Together, (2.17) and (2.18) prove that $v \in B_\infty^{s-1,\infty}$, and the proof of the theorem is complete.

4.3. Convergence estimates from L_1 to L_∞.

In this section we shall prove convergence estimates in the maximum-norm, with the smoothness of the data measured in L_1. We will show that for t bounded away from zero, we obtain $O(h^s)$ convergence in the maximum-norm (with $1 < s \le \mu$) when the initial function is in $B_1^{s,\infty}$ rather than in $B_\infty^{s,\infty}$, as was assumed in Theorem 1.3.

Theorem 3.1. Let $E(t)$ and $E_h(t)$ be as in Section 1 with P_h consistent with $-D^2$ and accurate of order μ, and let $1 < s \le \mu$. Then there exists a constant C such that for $v \in B_1^{s,\infty}$,

$$\|E_h(t)v - E(t)v\|_\infty \le Ct^{-1/2}h^s\|v\|_{B_1^{s,\infty}} \quad \text{for } t > 0.$$

Proof. Let $v_j = \mathcal{F}^{-1}(\phi_j\hat{v})$, $j \in Z$, and let f_τ be defined by (2.8), with $\tau = th^{-2}$. Then since $f_\tau(0) = 0$ and $\sum_j \phi_j(\xi) = 1$ for $\xi \neq 0$, we obtain for the error,

$$F_h(t)v = \mathcal{F}^{-1}\Big(\sum_{j=-\infty}^{\infty} \phi_j f_\tau(h\xi)\mathcal{F}\Big(\sum_{k=j-1}^{j+1} v_k\Big)\Big).$$

We notice that for $a \in L_1$,

$$\|\mathcal{F}^{-1}(a\hat{v})\|_\infty \le (2\pi)^{-1}\|a\hat{v}\|_1 \le (2\pi)^{-1}\|a\|_1\|\hat{v}\|_\infty \le (2\pi)^{-1}\|a\|_1\|v\|_1 ,$$

and hence using the definition of the seminorm $\|\ \|_{\dot{B}_1^{s,\infty}}$ that (cf. Section 2.6)

$$(3.1)\qquad \|F_h(t)v\|_\infty \le (2\pi)^{-1} \sum_{j=-\infty}^{\infty} \sum_{k=j-1}^{j+1} \|\phi_j f_\tau(h\cdot)\|_1 \|v_k\|_1$$

$$\le C\Big(\sum_{j=-\infty}^{\infty} \|\phi_j f_\tau(h\ \cdot)\|_1 2^{-js}\Big)\|v\|_{\dot{B}_1^{s,\infty}} .$$

We proceed to estimate $\|\phi_j f_\tau(h\cdot)\|_1$. Consider first the case $2^j h \le \pi/2$. We may write

$$f_\tau(h\xi) = \tau \int_0^1 \exp(-(1-\sigma)\tau(h\xi)^2 - \sigma\tau p(h\xi))b(h\xi)d\sigma ,$$

where $b(\xi) = \xi^2 - p(\xi)$. The accuracy of P_h and the parabolicity of the discrete problem then imply that for some $c > 0$ and for $|h\xi| \le \pi$,

$$|f_\tau(h\xi)| \le C\tau|h\xi|^{2+s} \exp(-c\tau(h\xi)^2) \le C|h\xi|^s \exp(-c\tau(h\xi)^2).$$

Hence,

$$(3.2)\qquad \|\phi_j f_\tau(h\xi)\|_1 \le C2^j(2^j h)^s \exp(-ct2^{2j}), \text{ for } 2^j h \le \pi/2.$$

In the case $2^j h \ge \pi/2$, we first notice that

$$(3.3)\qquad \|\phi_j \exp(-t\xi^2)\|_1 \le \int_{-\infty}^{\infty} \exp(-t\xi^2)d\xi = Ct^{-1/2}, \text{ for } t > 0.$$

Let $S_j = \{m \in Z : h2^{j-1} \le 2\pi m \le h2^{j+1}\}$. By periodicity and parabolicity we obtain

$$(3.4)\qquad \begin{aligned} \|\phi_j \exp(-\tau p(h\cdot))\|_1 &\le \sum_{m\in S_j} \int_{|h\xi-2\pi m|\le\pi} |\exp(-\tau p(h\xi - 2\pi m))|d\xi \\ &\le C2^j h \int_{|\xi h|\le\pi} \exp(-ct\xi^2)d\xi \le C2^j ht^{-1/2} . \end{aligned}$$

Hence by (3.3) and (3.4),

(3.5) $\|\phi_j f_\tau(h\cdot)\|_1 \le Ct^{-1/2}2^j h$, for $2^j h \ge \pi/2$.

The estimates (3.2) and (3.5) now give, with j_0 the largest integer such that $2^{j_0}h \le \pi/2$,

$$\sum_{-\infty}^{\infty} \|\phi_j f_\tau(h\cdot)\|_1 2^{-sj} \le C\{h^s \sum_{-\infty}^{j_0} 2^j \exp(-ct2^{2j}) + t^{-1/2} h \sum_{j_0+1}^{\infty} 2^{j(1-s)}\}$$

$$\le Ch^s t^{-1/2}\{\int_0^\infty \exp(-c\xi^2)d\xi + 1\} = Ch^s t^{-1/2} .$$

In view of (3.1) this proves the theorem.

As an example showing that Theorem 3.1 is an improvement over our previous Theorem 1.4 for functions of a type which is likely to appear in applications, consider the function H_s defined in Example II of Section 2.4. Recall that H_s is smooth outside the origin and that by Proposition 2.4.2, $H_s \in B_p^{r,\infty}$ if and only if $r \le s + 1/p$ for $1 \le p \le \infty$. The convergence estimate in the maximum norm given by Theorem 1.4 is therefore $O(h^s)$ for $0 < s \le \mu$ as h tends to zero, whereas our present Theorem 3.1 shows $O(h^{s+1})$ for $0 < s \le \mu-1$, which is best possible in view of Theorem 2.4. Notice also that Theorem 3.3.8, which was a simple consequence of the L_2 theory and the Sobolev inequality of Theorem 2.2.4, and did not take parabolicity into account, would give only $O(h^{s\mu/(\mu+2)})$ for $0 < s < \mu+2$.

4.4. Smoothing of initial data.

In the first three sections of this chapter we have investigated the relation between the rate of convergence of our semi-discrete approximation to the heat equation and the smoothness of the initial function. In particular we have seen that in order to obtain rapid convergence, even at a fixed positive time, considerable smoothness assumptions have to be imposed on the data. It is then natural to ask whether for non-smooth data, a preliminary smoothing could improve the convergence. It is our purpose in this section to show that this is the case.

We shall introduce a concept of a smoothing operator S_h $(0 < h \leq 1)$ which is suitable for our purposes by setting, for Φ a given function in M_∞,

$$S_h v = \mathcal{F}^{-1}(\Phi(h\xi)\hat{v}).$$

We call S_h a smoothing operator of orders (μ,ν) with μ and ν positive integers if there are two functions b_0 and b_1 in M_∞ such that

$$(4.1) \qquad \Phi(\xi) = 1 + \xi^\mu b_0(\xi), \text{ for } |\xi| \leq \pi,$$

$$(4.2) \qquad \Phi(\xi) = (\sin \tfrac{1}{2}\xi)^\nu b_1(\xi), \text{ for } |\xi| \geq \pi/2 .$$

The first condition means that in a certain sense S_h approximates the identity operator with order of accuracy μ, and the second condition makes Φ vanish of order ν for all non-zero multiples of 2π. Since

$$E_h(t)S_h v = \mathcal{F}^{-1}(\Phi(h\xi)\exp(-th^{-2}p(h\xi))\hat{v}),$$

the latter condition will dampen the frequences of v near the points where $p(h\xi)$ vanishes, that is the points for $\xi \neq 0$ where the periodic function $\exp(-th^{-2}p(h\xi))$ is not small.

We notice that a smoothing operator of orders (μ,ν) is also a smoothing operator of orders (μ',ν') for $\mu' \leq \mu$, $\nu' \leq \nu$.

The averaging operators

$$S_h^{(1)}v(x) = h^{-1}\int_{-h/2}^{h/2} v(x-y)dy$$

and

$$S_h^{(2)}v(x) = h^{-1}\int_{-h}^{h} (1 - |h^{-1}y|)v(x-y)dy ,$$

are examples of smoothing operators of orders (2,1) and 2,2), respectively. In fact,

$$S_h^{(j)}v = \mathcal{F}^{-1}(\Phi_j(h\xi)\hat{v}) ,$$

with

$$\Phi_j(\xi) = \left(\frac{\sin\frac{1}{2}\xi}{\frac{1}{2}\xi}\right)^j , \quad j = 1,2.$$

Obviously, $S_h^{(1)}$ is a bounded operator on L_∞ so that Φ_1 and $\Phi_2 = \Phi_1^2$ belong to M_∞. Since for $j = 1,2$,

$$\Phi_j(\xi) = 1 + O(\xi^2), \text{ as } \xi \to 0,$$

and since Φ_j is smooth, (4.1) holds with $\mu = 2$. Also, since, as can be seen for instance by the Carlson-Beurling inequality, ξ^{-j} is the restriction to $|\xi| \geq \pi/2$ of a function in M_∞, (4.2) holds with $\nu = j$. More generally, letting $q(\xi)$ be a polynomial such that

$$q(\sin \xi) = \xi^\mu + O(\xi^{2\mu}), \text{ as } \xi \to 0,$$

it is easy to see that the operator S_h defined by

$$\Phi(\xi) = \frac{q(\sin\frac{1}{2}\xi)}{(\frac{1}{2}\xi)^\mu} ,$$

is a smoothing operator of orders (μ,μ).

Our first result shows that for t bounded away from zero it is possible to obtain full convergence in L_p even if the initial function is only in L_p.

<u>Theorem 4.1</u>. Let $1 \leq p \leq \infty$, and let $E(t)$, $E_h(t)$ and P_h be as in Section 1, where in particular P_h has order of accuracy μ. Let S_h be a smoothing operator of orders (μ,μ). Then there exists a constant C such that for $v \in L_p$,

$$\|E_h(t)S_h v - E(t)v\|_p \leq Ct^{-\mu/2} h^\mu \|v\|_p, \text{ for } t > 0.$$

<u>Proof</u>. For $v \in \hat{C}_0^\infty$ we may write

(4.3) $$E_h(t)S_h v - E(t)v = \mathcal{F}^{-1}((\exp(-th^{-2}p(h\xi))\Phi(h\xi) - \exp(-t\xi^2))\hat{v}) .$$

Hence, setting $\tau = th^{-2}$, it is sufficient to prove that

$$(4.4)\qquad M_p(\exp(-\tau p(\xi))\Phi(\xi) - \exp(-\tau\xi^2)) \le C\tau^{-\mu/2},$$

and by Theorem 1.2.4, we may take $p = \infty$. Let $\chi \in C_0^\infty(-\pi,\pi)$ be such that $\chi = 1$ on $(-\pi/2,\pi/2)$. We may write

$$\exp(-\tau p)\Phi - \exp(-\tau\xi^2)$$

$$(4.5)\qquad = (\exp(-\tau p) - \exp(-\tau\xi^2))\chi\Phi + \exp(-\tau\xi^2)(\Phi-1)\chi - \exp(-\tau\xi^2)(1-\chi) + \exp(-\tau p)\Phi(1-\chi)$$

$$= \sum_{l=1}^{4} \sigma_l(\xi),$$

and we shall prove that each of the σ_l can be estimated in the M_∞-norm by $C\tau^{-\mu/2}$.

Let again $f_\tau(\xi) = \exp(-\tau p(\xi)) - \exp(-\tau\xi^2)$. Then by (1.13) with $\alpha = 0$ ($m_j = M_\infty(\phi_j(h^{-1}\cdot)f_\tau)$ in the present notation) and with j_0 the largest integer such that $2^{j_0}h \le \pi$, we obtain for the first term

$$M_\infty(\sigma_1) = M_\infty(f_\tau\chi\Phi) \le M_\infty(\Phi\chi) \sum_{-\infty}^{j_0} M_\infty(f_\tau\phi_j(h^{-1}\cdot))$$

$$\le C \sum_{-\infty}^{j_0} \tau(2^jh)^{\mu+2}\exp(-c\tau(2^jh)^2) \le C\tau^{-\mu/2}.$$

By (4.1) we obtain for the second term in (4.5)

$$M_\infty(\sigma_2) \le M_\infty(\chi b_0)M_\infty(\xi^\mu \exp(-\tau\xi^2)) \le C\tau^{-\mu/2}.$$

To estimate the third term, we obtain similarly, since $\xi^{-\mu}(1-\chi(\xi)) \in M_\infty$ (e.g. by the Carlson-Beurling inequality),

$$M_\infty(\sigma_3) \le M_\infty(\xi^{-\mu}(1-\chi))M_\infty(\xi^\mu\exp(-\tau\xi^2)) \le C\tau^{-\mu/2}.$$

Finally, for the fourth term we obtain by (4.2) (with $\nu = \mu$) using in the last step Theorem 1.4.1 and (1.5) (with $\alpha = \mu$),

$$M_\infty(\sigma_4) \le M_\infty((\sin \tfrac{1}{2}\xi)^\mu\exp(-\tau p))M_\infty(b_1(1-\chi)) \le CM_\infty((e^{i\xi} - 1)^\mu\exp(-\tau p)) \le C\tau^{-\mu/2}.$$

Together these four estimates complete the proof of (4.4) with $p = \infty$ and hence of the theorem.

Employing the general approach of Section 3, we shall now give an estimate for the rate of convergence in the maximum-norm when the smoothness of the data is measured in L_1, in the presence of a smoothing operator. For simplicity, we shall estimate the rate of convergence for t bounded away from zero.

<u>Theorem 4.2</u>. Let $E(t)$, $E_h(t)$ and P_h be as above, and let S_h denote a smoothing operator of orders (μ,ν). Assume that $s \geq \mu-\nu$ and that $1 < s \leq \mu$. Then for each $t_0 > 0$ there exists a constant C such that for $v \in B_1^{s,\infty}$,

$$\|E_h(t)S_h v - E(t)v\|_\infty \leq Ch^\mu \|v\|_{B_1^{s,\infty}}, \quad \text{for } t \geq t_0.$$

<u>Proof</u>. We have as before (4.3) so that (cf. (3.1)),

$$\|E_h(t)S_h v - E(t)v\|_\infty \leq C\Big(\sum_{-\infty}^{\infty} m_j 2^{-sj}\Big)\|v\|_{\dot{B}_1^{s,\infty}}, \tag{4.6}$$

where

$$m_j = \|\phi_j(\exp(-th^{-2}p(h\xi))\Phi(h\xi) - \exp(-t\xi^2))\|_1 .$$

Setting again $\tau = th^{-2}$ we shall once more use (4.5) and estimate the L_1-norms of $\phi_j\sigma_l(h\xi)$, $l = 1,2,3,4$. For σ_1 we obtain by (3.2) (with the same f_τ) with $s = \mu$,

$$\|\phi_j\sigma_1(h\cdot)\|_1 \leq \|\chi\Phi\|_\infty \|\phi_j f_\tau(h\cdot)\|_1 \leq C2^{j(1+\mu)}h^\mu \exp(-ct2^{2j}).$$

Similarly, using (4.1) we obtain

$$\|\phi_j\sigma_2(h\cdot)\|_1 \leq C\int \phi_j(\xi)|h\xi|^\mu \exp(-t\xi^2)d\xi \leq C2^{j(1+\mu)}h^\mu \exp(-ct2^{2j}).$$

For σ_3 we find, since $|\xi|^{-\nu}(1-\chi(\xi))$ is bounded,

$$\|\phi_j\sigma_3(h\cdot)\|_1 \leq C\int \phi_j(\xi)|h\xi|^\nu \exp(-t\xi^2)d\xi \leq Ct^{-(1+\nu)/2}h^\nu ,$$

and finally, for σ_4 we obtain as in (3.4), with that notation,

$$\|\phi_j\sigma_4(h\cdot)\|_1 \le C \int \phi_j(\xi)(\sin \tfrac{1}{2}h\xi)^\nu \exp(-c\tau p(h\xi))d\xi$$

$$\le C \sum_{m\in S_j} \int_{|h\xi|\le\pi} |h\xi|^\nu \exp(-ct\xi^2)d\xi \le Ct^{-(1+\nu)/2} h^{\nu+1} 2^j .$$

Together these four estimates imply, with j_0 the largest integer such that $2^{j_0}h \le \pi/2$, since $s \ge \mu-\nu$ and $t \ge t_0 > 0$,

$$\sum_{-\infty}^{\infty} m_j 2^{-sj} \le C \sum_{j=-\infty}^{\infty} \Big(\sum_{l=1}^{4} \|\phi_j\sigma_l(h\cdot)\|_1\Big)2^{-js}$$

$$\le Ch^\mu \sum_{-\infty}^{j_0} 2^{j(1+\mu-s)}\exp(-ct2^{2j}) + Ct_0^{-(1+\nu)/2} h^\nu \sum_{j_0}^{\infty} (2^{-js} + h2^{j(1-s)}) \le Ch^\mu .$$

By (4.6) this completes the proof of the theorem.

As a special case of Theorem 4.2, it follows that if the initial function is the function $H_{s-1} \in B_1^{s,\infty}$ of Example II of Section 2.4 with $s > 1$ but close to 1, and if the smoothing operator $S_h^{(1)}$ of orders (2,1) is applied, then for a second order accurate semi-discretization of the heat equation, the convergence rate in the maximum norm is $O(h^2)$. In order to obtain the same convergence estimate by Theorem 4.1 it would have been necessary to employ for instance, the somewhat more complicated smoothing operator $S_h^{(2)}$ of orders (2,2).

References.

Section 1 is based on the theory of difference approximations to parabolic problems, developed in [1], [3], [4], [5], [7] and references. Theorem 2.1 was proved in [1], and Theorems 2.2 and 2.3 are contained in [3]. Theorem 2.4 and the material of Section 3 are taken from [6] and Section 4 is compiled from [2] and [6]. Several of the papers quoted treat parabolic systems of arbitrary order with variable coefficients.

1. G.W. Hedstrom, The rate of convergence of some difference schemes, SIAM J. Numer. Anal. 5 (1968), 363-406.

2. H.O. Kreiss, V. Thomée and O.B. Widlund, Smoothing of initial data and rates of convergence for parabolic difference equations, Comm. Pure Appl. Math. 23 (1970), 241-259.

3. J. Löfström, Besov spaces in theory of approximation, Ann. Mat. Pura Appl. 85 (1970), 93-184.

4. J. Peetre and V. Thomée, On the rate of convergence for discrete initial value problems, Math. Scand. 21 (1967), 159-176.

5. V. Thomée, Parabolic difference operators, Math. Scand. 19 (1966), 77-107.

6. V. Thomée and L. Wahlbin, Convergence rates of parabolic difference schemes for non-smooth data, Math. Comp. 28 (1974), 1-13.

7. O.B. Widlund, On the rate of convergence for parabolic difference schemes II, Comm. Pure Appl. Math. 23 (1970), 79-96.

CHAPTER 5. FIRST ORDER HYPERBOLIC EQUATIONS.

In this chapter we shall be concerned with stability and convergence results in L_p with $p \neq 2$ for difference approximations to first order hyperbolic equations. We first show in Section 1 that the initial value problem for a symmetric hyperbolic system is well posed in L_p for $p \neq 2$ if and only if the system, modulo a unitary transformation of the dependent variables, is a system of uncoupled first order scalar differential equations. We shall then restrict our attention in the rest of the chapter to a single equation in one space dimension. In Section 2 we find necessary and sufficient conditions for stability in L_p, $p \neq 2$, of a class of difference operators E_k consistent with such an equation and in Section 3 we determine the rate of growth of E_k^n in some non-stable situations. In Section 4 we give precise estimates in terms of the smoothness of the initial data for the rate of convergence of both stable and unstable difference approximations. Finally, in Section 5 we consider a second order accurate finite difference scheme based on the Lax-Wendroff operator for a simple semi-linear equation. Here the analysis takes place in the Besov space $B_2^{1/2,1}$ which is seen to be appropriate for the problem.

5.1. The initial value problem for a symmetric hyperbolic system in L_p.

We shall consider the initial value problem

$$(1.1)\qquad \begin{aligned} \frac{\partial u}{\partial t} &= \sum_{j=1}^{d} A_j \frac{\partial u}{\partial x_j}, \quad \text{for} \quad x \in R^d,\ t > 0, \\ u(x,0) &= v(x), \end{aligned}$$

where $A_1, \ldots A_d$ are constant hermitean $N \times N$ matrices and $u = u(x,t)$ and $v = v(x)$ are N-vector valued functions.

For N-vectors $v = (v_1,\dots,v_N)'$ and $w = (w_1,\dots,w_N)'$, where v' denotes the transpose of v, we use the scalar product $\langle v,w\rangle = \sum_{j=1}^{N} v_j\bar{w}_j$ and the corresponding Euclidean norm $|v| = \langle v,v\rangle^{1/2}$, and for $N \times N$ matrices A the associated matrix norm $|A| = \sup\{|Av| : |v| = 1\}$. With the corresponding definition of L_p norms of vector and matrix valued functions, and taking Fourier transforms element-wise we may define as in Section 1.1, for a a $N \times K$ matrix valued function on R^d,

$$M_p(a) = \sup\{\|\mathcal{F}^{-1}(a\hat{v})\|_p : v = (v_1,\dots,v_K)' \in \hat{C}_0^\infty, \|v\|_p \le 1\}, \tag{1.2}$$

and we say that $a \in M_p$ if this norm is finite. If $a = (a_{jk})$ we have $a \in M_p$ if and only if $a_{jk} \in M_p$ for all j and k, and a norm equivalent to the one in (1.2) is $\max_{j,k} M_p(a_{jk})$. Several of the results of Chapter 1 hold in the present context after obvious modifications. In particular, if $a,b \in M_p$ and a is a $M \times K$ matrix and b a $K \times N$ matrix, then $ab \in M_p$ and

$$M_p(ab) \le M_p(a)M_p(b). \tag{1.3}$$

We now define the solution operator $E(t)$ of the problem (1.1) by (cf. Section 3.1),

$$E(t)v = \mathcal{F}^{-1}(\exp(t\hat{P})\hat{v}), \text{ for } v = (v_1,\dots,v_N)' \in \hat{C}_0^\infty,$$

where

$$\hat{P}(\xi) = i \sum_{j=1}^{d} A_j\xi_j .$$

We say that (1.1) is well posed in L_p if for each $T > 0$ there exists a constant C such that

$$\|E(t)v\|_p \le C\|v\|_p \quad \text{for } 0 \le t \le T,\ v \in \hat{C}_0^\infty.$$

Since $\sum_{j=1}^{d} A_j\xi_j$ is hermitean, $\exp(t\hat{P})$ is unitary, so that

$$M_2(\exp(t\hat{P})) = \sup_\xi |\exp(t\hat{P}(\xi))| = 1,$$

and hence by Proposition 3.1.1, (1.1) is well posed in L_2. However, it is the main result of this section that for $p \neq 2$, well-posedness is an exception.

Theorem 1.1. Let $1 \leq p \leq \infty$, $p \neq 2$. Then the initial value problem (1.1) is well posed in L_p if and only if the matrices $A_1,\dots,A_d$ commute.

Proof. As is well known, (and also follows from the proof of Lemma 1.2 below) the commutativity of $A_1,\dots,A_d$ implies that the A_j may be simultaneously diagonalized by a unitary matrix U so that $D_j = UA_jU^*$ (U^* denotes the conjugate transpose of U) are real diagonal matrices for $j = 1,\dots,d$. Notice that this means that if we introduce $\tilde{u} = Uu$ as a new variable in (1.1), the system takes the form

$$\frac{\partial\tilde{u}}{\partial t} = \sum_{j=1}^{d} D_j \frac{\partial\tilde{u}}{\partial x_j},$$

that is, it is a system of N uncoupled first order differential equations.

Assume now that the A_j commute. With U as above we have then

$$\exp(t\hat{P}(\xi)) = \exp(ti \sum_{j=1}^{d} U^*D_jU\xi_j) = U^*\exp(it \sum_{j=1}^{d} D_j\xi_j)U,$$

and since the latter exponential matrix is a diagonal matrix, with elements of the form $\exp(itd(\xi))$ where $d(\xi)$ is a real linear function, it follows that $\exp(t\hat{P})$ is in M_p for $1 \leq p \leq \infty$. This proves the sufficiency part of the theorem.

The main steps in the proof of the necessity part of Theorem 1.1 will be carried out in two lemmas.

Lemma 1.1. Let $1 \leq p \leq \infty$, $p \neq 2$. If $\exp(\hat{P}) \in M_p$, then the eigenvalues of $A(\xi) = -i\hat{P}(\xi) = \sum_{j=1}^{d} A_j\xi_j$ can be chosen as real linear functions of ξ.

Proof. Let B be an open ball in R^d such that the eigenvalues and the corresponding eigenvectors of $A(\xi)$ can be chosen as C^∞ functions on B. We shall prove below that these eigenvalues are of the form

(1.4) $\lambda_k(\xi) = \lambda_0^{(k)} + \sum_{j=1}^{d} \lambda_j^{(k)}\xi_j$, $k = 1,\ldots,N$, for $\xi \in B$.

It follows then that for complex z and $\xi \in B$,

$$\det(zI - A(\xi)) = \prod_{k=1}^{N} (z - \lambda_0^{(k)} - \sum_{j=1}^{d} \lambda_j^{(k)}\xi_j).$$

Since both sides are entire analytic functions in ξ, this equality holds for all complex z and all real (and even all complex) ξ. In particular, $\lambda_0^{(k)} = 0$ by homogeneity, and the then linear functions $\lambda_k(\xi)$, $k = 1,\ldots,N$, are the eigenvalues of $A(\xi)$ for all $\xi \in R^d$. This proves the lemma, assuming that (1.4) holds.

In order to prove (1.4), let $\lambda(\xi) \in C^\infty(B)$ be an eigenvalue of $A(\xi)$ in B, and let $v(\xi) \in C^\infty(B)$ be a corresponding eigenvector. We shall prove that at an arbitrary point $\xi^0 \in B$, all the second order derivatives of $\lambda(\xi)$, $\lambda_{jk}^0 = \partial^2\lambda(\xi^0)/\partial\xi_j\partial\xi_k$ vanish. For this purpose, let $w \in C_0^\infty(R^d)$ be such that $\langle v,w\rangle = w^*v = 1$ on B, and choose χ in $C_0^\infty(B)$ with $\chi(\xi^0) = 1$. Then

$$\chi(\xi)\exp(in\lambda(\xi))v(\xi) = \chi(\xi)\exp(n\hat{P}(\xi))v(\xi)$$

and hence

$$\chi(\xi)\exp(in\lambda(\xi)) = \langle\chi(\xi)\exp(n\hat{P}(\xi))v(\xi),w(\xi)\rangle .$$

Using (1.3) and Theorem 1.2.8 we obtain since χv and w both belong to $C_0^\infty(R^d)$ and hence to M_p,

(1.5) $M_p(\chi \exp(in\lambda)) \leq M_p(\exp(n\hat{P}))M_p(\chi v)M_p(w^*) = CM_p(\exp(\hat{P})) = C$, $n = 1,2,\ldots$.

Setting

$$\tilde{\lambda}(\xi) = \lambda(\xi+\xi^0) - \lambda(\xi^0) - \langle\xi,\text{grad } \lambda(\xi^0)\rangle,$$

and

$$h_n(\xi) = \chi(\xi^0 + n^{-1/2}\xi)\exp(in\tilde{\lambda}(n^{-1/2}\xi)) ,$$

we find by Theorem 1.2.8 and (1.5) that $M_p(h_n) \leq C$, for $n = 1,2,\ldots$. On the other hand, we have, uniformly on compact subsets of R^d,

$$\lim_{n\to\infty} h_n(\xi) = \exp(iQ(\xi)), \text{ where } Q(\xi) = \frac{1}{2}\sum_{j,k=1}^{d} \lambda^0_{jk}\xi_j\xi_k .$$

By Theorem 1.2.6 we conclude that $\exp(iQ) \in M_p$, and since $p \neq 2$ this implies by Corollary 1.5.3 that Q vanishes identically. Hence the $\lambda^0_{jk} = \partial^2\lambda(\xi^0)/\partial\xi_j\partial\xi_k$ vanish, which completes the proof of the lemma.

We next prove that the fact that the eigenvalues of $A(\xi) = i\hat{P}(\xi)$ can be chosen as linear functions implies that the A_j commute.

Lemma 1.2. Let $A_1,\ldots,A_d$ be hermitean $N \times N$ matrices and assume that the eigenvalues of $A(\xi) = \sum_{j=1}^{d} A_j\xi_j$ can be chosen as real linear functions of ξ on R^d. Then the matrices $A_1,\ldots,A_d$ commute.

Proof. By the spectral theorem, a hermitean matrix A with r distinct eigenvalues can be represented as

$$A = \sum_{j=1}^{r} \lambda_j E_j ,$$

where E_j are mutually orthogonal hermitean projections given by

$$E_j = \prod_{\substack{k=1\\k\neq j}}^{r} \frac{A-\lambda_k I}{\lambda_j-\lambda_k} .$$

Let $\lambda_j(\xi)$, $j = 1,\ldots,r$, be the distinct linear functions which constitute the set of eigenvalues of $A(\xi)$ for $\xi \in R^d$. Then, except for ξ in the set V where two or more $\lambda_j(\xi)$ coincide, we have

$$A(\xi) = \sum_{j=1}^{r} \lambda_j(\xi)E_j(\xi) ,$$

where the mutually orthogonal hermitean projections $E_j(\xi)$ are given by

$$E_j(\xi) = F_j(\xi) / \prod_{k \neq j} (\lambda_j(\xi) - \lambda_k(\xi)), \text{ where } F_j(\xi) = \prod_{k \neq j} (A(\xi) - \lambda_k(\xi)I).$$

We shall prove that $E_j(\xi)$ is a constant matrix for $\xi \notin V$. We notice first that for $k \neq j$ the linear function $\lambda_j(\xi) - \lambda_k(\xi)$ is a factor in each element of the polynomial $F_j(\xi)$. In fact, since

$$F_j(\xi) = \prod_{k \neq j} (\lambda_j(\xi) - \lambda_k(\xi)) E_j(\xi) ,$$

and $|E_j(\xi)| = 1$ for $\xi \notin V$ we obtain in the limit that $F_j(\xi) = 0$ for $\lambda_j(\xi) - \lambda_k(\xi) = 0$. We may hence successively remove all the $r-1$ linear factors $\lambda_j(\xi) - \lambda_k(\xi)$, $k \neq j$, from $F_j(\xi)$. Since $F_j(\xi)$ has degree $r-1$, it follows that $E_j(\xi)$ is constant for $\xi \notin V$ and hence for these ξ,

$$A(\xi) = \sum_{j=1}^{r} \lambda_j(\xi) E_j .$$

By continuity this relation then holds for all ξ on R^d. Since the E_j commute and $A_k = A(e_k)$, where e_k denotes the kth unit vector, this completes the proof of the lemma.

We can now complete the proof of Theorem 1.1. In fact, by Proposition 3.1.1, the well-posedness of (1.1) implies that $\exp(\hat{P}) \in M_p$, and by Lemmas 1.1 and 1.2, we may hence conclude that $A_1, \ldots, A_d$ commute.

5.2. Stability in L_p of difference analogues of $\partial u/\partial t = \partial u/\partial x$.

In this and the following two sections we shall be concerned with completely discrete finite difference approximations to the initial value problem

$$\frac{\partial u}{\partial t} = \frac{\partial u}{\partial x}, \text{ for } x \in R,\ t > 0, \tag{2.1}$$

$$u(x,0) = v(x).$$

This problem has the exact solution

$$u(x,t) = E(t)v(x) = v(x+t) = \mathcal{F}^{-1}(e^{i\xi t}\hat{v})(x)$$

and is well posed in L_p for $1 \leq p \leq \infty$.

The finite difference operators will be of the form (cf. Section 3.2)

$$(2.2) \qquad E_k v(x) = \mathcal{F}^{-1}(\hat{E}_k \hat{v})(x) = \mathcal{F}^{-1}(e(h\cdot)\hat{v})(x) = \sum_{j=-\infty}^{\infty} a_j v(x+hj), \; k/h = \lambda = \text{constant},$$

with a characteristic function

$$e(\xi) = \hat{E}_k(h^{-1}\xi) = \sum_{j=-\infty}^{\infty} a_j e^{ij\xi}$$

which is a rational trigonometric function independent of h. (It is in fact sufficient below to assume $e(\xi)$ real analytic.) If E_k is consistent with (2.1) we have

$$e(\xi) = \exp(i\lambda\xi + o(\xi)) \quad \text{as} \quad \xi \to 0.$$

Further, E_k is stable in L_2 if and only if

$$(2.3) \qquad |e(\xi)| \leq 1 \quad \text{for} \quad \xi \in R.$$

In this section we shall give necessary and sufficient conditions for an L_2-stable operator E_k of the form (2.2) to be stable also in L_p for $p \neq 2$. In the next section we shall then estimate the rate of growth in the unstable case, and in Section 4 we give convergence estimates for both stable and unstable operators.

The following is the main result of this section. We phrase it in such a way as to make it apparent that it is in fact a result concerning operators of the form (2.2) which is independent of their relation to (2.1).

<u>Theorem 2.1</u>. Let $1 \leq p \leq \infty$, $p \neq 2$, and assume that E_k is an operator of the form (2.2).

Then E_k is stable in L_p if and only if one of the following two conditions is satisfied:

(i) There exist constants c and α, with $|c| = 1$ and α real, such that $e(\xi) = ce^{i\alpha\xi}$.

(ii) $|e(\xi)| < 1$ except for at at most a finite number of points ξ_q, $q = 1,\dots,Q$, in $[-\pi,\pi]$, where $|e(\xi)| = 1$. For $q = 1,\dots,Q$ there are constants α_q, β_q, and ν_q, with α_q real, Re $\beta_q > 0$, and ν_q an even natural number, such that

$$e(\xi_q+\xi) = e(\xi_q)\exp(i\alpha_q\xi - \beta_q\xi^{\nu_q}(1+o(1))), \text{ as } \xi \to 0.$$

<u>Proof</u>. We first show that stability implies that (i) or (ii) is satisfied.

Notice that by Theorem 1.2.4, (2.3) is necessary for stability also in L_p. Since $e(\xi)$ is 2π-periodic and analytic, it follows that one of the following two conditions holds:

(i') $|e(\xi)| = 1$ for all real ξ,

(ii') $|e(\xi)| < 1$ for all but at most a finite number of points ξ_q, $q = 1,\dots,Q$, in $[-\pi,\pi]$.

We shall show that if E_k is stable in L_p, $p \neq 2$, then (i') implies (i) and (ii') implies (ii). If this were not so, it would be possible to find a real ξ_0 such that $|e(\xi_0)| = 1$ and α,β, and ν with α and β real, $\beta \neq 0$ and ν an integer > 1 such that

(2.4) $$e(\xi_0+\xi) = e(\xi_0)\exp(i\alpha\xi + i\beta\xi^{\nu}(1+o(1))), \text{ as } \xi \to 0.$$

Let $a(\xi) = e(\xi_0)^{-1}e(\xi_0+\xi)e^{-i\alpha\xi}$. By the assumed L_p-stability we have

(2.5) $$M_p(a^n) = M_p(e^n) \le C, \; n = 1,2,\dots .$$

From (2.4) we find that $\lim_{n\to\infty} a^n(n^{-1/\nu}\xi) = \exp(i\beta\xi^{\nu})$, uniformly on compact sets. Theorem 1.2.6 and (2.5) then prove that $\exp(i\beta\xi^{\nu}) \in M_p$, which contradicts Corollary 1.5.3. This shows the necessity of (i) or (ii) for L_p-stability.

We now turn to the proof of the sufficiency of (i) or (ii). In the case (i) this is obvious, since E_k^n is then a translation operator.

It remains to consider the case (ii). By Theorem 1.2.4 it suffices to treat $p = \infty$, and by Theorem 1.4.1 it is enough to prove that

(2.6) $$M_\infty(\eta e^n) \le C, \quad n = 1,2,\dots ,$$

where $\eta \in C_0^\infty$, $\eta = 1$ on $(-\pi-\varepsilon,\pi+\varepsilon)$ and $\eta = 0$ outside $(-\pi-2\varepsilon,\pi+2\varepsilon)$, for some $\varepsilon \in (0,\pi/8)$.

Let $\delta > 0$ be smaller than the distance between the ξ_q:s, and let χ be a C^∞ function with $|\chi| \le 1$ such that

$$\chi(\xi) = \begin{cases} 1 & \text{for } |\xi| \le \delta/4, \\ 0 & \text{for } |\xi| > \delta/2 . \end{cases}$$

Set $\chi_q(\xi) = \eta(\xi)\chi(\xi-\xi_q)$, $q = 1,\dots,Q$. Without loss of generality we may assume that $4\varepsilon < \delta$. Then the only points in $\operatorname{supp}(\eta)$ where $|e(\xi)| = 1$ are ξ_q, $q = 1,\dots,Q$. Setting $\chi_0(\xi) = \eta(\xi) - \sum_{q=1}^{Q} \chi_q(\xi)$ we obtain

(2.7) $$M_\infty(\eta e^n) \le \sum_{q=0}^{Q} M_\infty(\chi_q e^n) .$$

We first estimate $M_\infty(\chi_0 e^n)$. Since $|e(\xi)| \le \kappa < 1$ on the support of χ_0, we have

$$\|\chi_0 e^n\|_2 \le C\kappa^n ,$$

and

$$\left\|\frac{d}{d\xi} (\chi_0 e^n)\right\|_2 \le \|\chi_0' e^n\|_2 + n\|\chi_0 e' e^{n-1}\|_2 \le Cn\kappa^n,$$

so that by the Carlson-Beurling inequality,

(2.8) $$M_\infty(\chi_0 e^n) \le Cn^{1/2}\kappa^n \le C, \quad n = 1,2,\dots .$$

We now turn to the remaining q:s. Letting $a(\xi) = e(\xi_q)^{-1}e(\xi_q+\xi)\exp(-i\alpha\xi_q)$ we have

$$M_\infty(\chi_q e^n) = M_\infty(\chi a^n) .$$

Our assumption implies that (dropping the index q) with $\operatorname{Re}\beta > 0$ and ν even,

$$a(\xi) = \exp(-\beta\xi^{\nu}(1+o(1))), \text{ as } \xi \to 0.$$

For δ sufficiently small we then have for $|\xi| \le \delta/2$, and some $c > 0$,

$$|a(\xi)| \le \exp(-c\xi^{\nu}),$$

$$\left|\frac{d}{d\xi} a(\xi)\right| \le C|\xi|^{\nu-1}\exp(-c\xi^{\nu}).$$

Hence

$$\|\chi a^n\|_2 \le \left(\int_{-\infty}^{\infty} \exp(-2cn\xi^{\nu})d\xi\right)^{1/2} = Cn^{-1/(2\nu)}, \text{ for } n = 1,2,\ldots$$

and

$$\left\|\frac{d}{d\xi}(\chi a^n)\right\|_2 \le \|\chi' a^n\|_2 + n\|\chi a' a^{n-1}\|_2$$

$$\le Cn^{-1/(2\nu)} + Cn\left(\int_{-\infty}^{\infty} |\xi|^{2(\nu-1)}\exp(-2cn\xi^{\nu})d\xi\right)^{1/2} \le Cn^{1/(2\nu)}.$$

The Carlson-Beurling inequality then gives

$$M_{\infty}(\chi a^n) \le C, \; n = 1,2,\ldots .$$

Together with (2.7) and (2.8), this proves (2.6) and completes the proof of the theorem.

If E_k is consistent with (2.1) and L_p-stable for $p \neq 2$, then in the case (i) of the theorem, $c = 1$ and $\alpha = \lambda$, and E_k is identical to the exact solution operator $E(k)$. In case (ii), $\xi = 0$ has to be one of the points for which $|e(\xi)| = 1$ and at that point, $\xi = \xi_1$, say, we have $e(\xi_1) = 1$, $\alpha_1 = \lambda$, and $\nu_1 = \mu+1$ where μ is the order of accuracy of E_k. Since ν_1 is an even number, only operators of odd order of accuracy can be L_p-stable for $p \neq 2$.

As examples we consider the simple operator,

$$E_k^{(1)}v(x) = \lambda v(x+h) + (1-\lambda)v(x),$$

obtained by replacing derivatives in (2.1) by forward difference quotients, the Lax-Wendroff operator

$$E_k^{(2)}v(x) = \frac{1}{2}(\lambda^2+\lambda)v(x+h) + (1-\lambda^2)v(x) + \frac{1}{2}(\lambda^2-\lambda)v(x-h),$$

and the implicit "box" operator defined by

$$(1-\lambda)E_k^{(3)}v(x+h) + (1+\lambda)E_k^{(3)}v(x) = (1+\lambda)v(x+h) + (1-\lambda)v(x).$$

These operators have the characteristic functions

$$e_1(\xi) = 1+\lambda(e^{i\xi}-1)$$

$$e_2(\xi) = 1-\lambda^2(1-\cos\xi)+i\lambda\sin\xi,$$

$$e_3(\xi) = \frac{e^{i\xi}+\kappa}{\kappa e^{i\xi}+1}, \text{ where } \kappa = \frac{1-\lambda}{1+\lambda}.$$

It is obvious directly from the definition that $E_k^{(1)}$ is stable in L_p for $1 \le p \le \infty$ when $\lambda \le 1$. For $\lambda = 1$ we have $e_1(\xi) = e^{i\xi}$ and for $\lambda < 1$, $|e_1(\xi)| < 1$ for $0 < |\xi| \le \pi$, and

$$e_1(\xi) = \exp(i\lambda\xi - \frac{1}{2}\lambda(1-\lambda)\xi^2(1+o(1))), \text{ as } \xi \to 0,$$

which agrees with the conclusion of Theorem 2.1. For $E_k^{(2)}$ an easy calculation shows

$$|e_2(\xi)|^2 = 1-\lambda^2(1-\lambda^2)(1-\cos\xi)^2 \le 1, \text{ for } \lambda \le 1.$$

For $\lambda = 1$ we have again $e_2(\xi) = e^{i\xi}$ and for $\lambda < 1$, $|e_2(\xi)| < 1$ for $0 < |\xi| \le \pi$, and

$$e_2(\xi) = \exp(i\lambda\xi + i\frac{\lambda}{6}(\lambda^2-1)\xi^3(1+o(1))), \text{ as } \xi \to 0.$$

It follows by Theorem 2.1 that $E_k^{(2)}$ is unstable in L_p for $p \neq 2$, when $\lambda < 1$. For the operator $E_k^{(3)}$, finally, $|e_3(\xi)| \equiv 1$ for any choice of λ. Except for $\lambda = 1$, when $e_3(\xi) = e^{i\xi}$, $E_k^{(3)}$ is hence unstable in L_p for $p \neq 2$.

5.3. Growth in the unstable case.

In this section we shall study the growth with n of the operator norm

$$\|E_k^n\|_p = \sup\{\|E_k^n v\|_p : \|v\|_p = 1\} = M_p(e^n),$$

in unstable situations, where E_k is an L_2-stable operator of the form (2.2) with characteristic function $e(\xi)$. For simplicity we consider only the cases

(3.1) $\quad |e(\xi)| \equiv 1 \quad \text{for} \quad \xi \in R,$

and

(3.2) $\quad |e(\xi)| < 1, \text{ for } 0 < |\xi| \leq \pi \quad \text{with} \quad e(0) = 1.$

In the following, let $\tilde{p} = |1/2 - 1/p|$.

We first assume that (3.1) holds.

Theorem 3.1. Let $1 \leq p \leq \infty$, and assume that E_k is unstable in L_p for $p \neq 2$, and that its characteristic function satisfies (3.1). Then there exist positive constants c and C such that

(3.3) $\quad cn^{\tilde{p}} \leq \|E_k^n\|_p \leq Cn^{\tilde{p}}, \text{ for } n = 1,2,\ldots .$

Proof. We may write $e(\xi) = \exp(i\theta(\xi))$ with θ real. Since E_k is unstable in L_p, for $p \neq 2$ we may assume by Theorem 2.1 that $\theta \in C^2(I)$ and $\theta'' \not\equiv 0$ on I, for some open interval I. We first establish the upper bound in (3.3). With η as in Theorem 1.4.1 we have

$$\|\eta e^n\|_2 \leq C, \quad \|(\eta e^n)'\|_2 \leq Cn,$$

so that by Theorem 1.4.1 and the Carlson-Beurling inequality

$$M_\infty(e^n) \leq CM_\infty(\eta e^n) \leq Cn^{1/2},$$

which proves the estimate from above in (3.3) for $p = \infty$. Since $M_2(e^n) = 1$, the result now follows for general p by Theorem 1.2.5.

We now turn to the estimate from below. If $\chi \in C_0^\infty(I)$ then by Corollary 1.5.1, for some $c > 0$,

$$cn^{\tilde{p}} \leq M_p(\chi \exp(in\theta(\xi))) = M_p(\chi e^n) \leq M_p(\chi)M_p(e^n), \text{ for } n = 1,2,\ldots .$$

This proves the estimate from below in (3.3) and hence completes the proof of the theorem.

The implicit operator $E_k^{(3)}$ defined in the preceeding section is an example of the situation in Theorem 3.1 for $\lambda \neq 1$.

We next consider the case when (3.2) holds. Since we shall assume non-stability, we may write, by Theorem 2.1,

$$e(\xi) = \exp(i\alpha\xi + i\beta\xi^\nu(1+o(1))), \text{ as } \xi \to 0, \tag{3.4}$$

$$|e(\xi)| = \exp(-\gamma\xi^\sigma(1+o(1))), \text{ as } \xi \to 0, \tag{3.5}$$

$$|e(\xi)| \leq \exp(-\gamma'\xi^\sigma), \text{ for } |\xi| \leq \pi, \tag{3.6}$$

where α and β are real, $\beta \neq 0$, γ and γ' positive and $\nu < \sigma$, σ even. When E_k is consistent with (1.1), we have $\alpha = \lambda$ and E_k is accurate of order $\nu-1$. The number σ in (3.6) is referred to as the order of dissipation of E_k.

Theorem 3.2. Let $1 \leq p \leq \infty$, and assume that (3.4), (3.5) and (3.6) hold. Then there exist positive constants c and C such that

$$cn^{\tilde{p}(1-\nu/\sigma)} \leq \|E_k^n\|_p \leq Cn^{\tilde{p}(1-\nu/\sigma)}, \text{ for } n = 1,2,\ldots .$$

Proof. We start with the estimate from above. Let $a(\xi) = e(\xi)e^{-i\alpha\xi}$. With η as in Theorem 1.4.1 we have this time by (3.6),

$$\|\eta a^n\|_2 \leq \left(\int_{-\infty}^{\infty} \exp(-2\gamma' n\xi^\sigma)d\xi\right)^{1/2} = Cn^{-1/(2\sigma)},$$

and using also (3.4),

$$\|(\eta a^n)'\|_2 \le Cn\left(\int_{-\infty}^{\infty} |\xi|^{2(\nu-1)} \exp(-2\gamma' n\xi^\sigma)d\xi\right)^{1/2} + Cn^{-1/(2\sigma)}$$

$$\le Cn^{1-\nu/\sigma+1/(2\sigma)} .$$

Hence the Carlson-Beurling inequality gives

$$M_\infty(\eta a^n) \le Cn^{(1-\nu/\sigma)/2},$$

which by Theorem 1.4.1 proves the result for $p = \infty$. Since $M_2(e^n) = 1$, the result follows for general p by Theorem 1.2.5.

We now turn to the estimate from below. We may assume as before that $p \ge 2$. Let $\chi \in C_0^\infty$, with $\chi \not\equiv 0$ and $0 \notin \operatorname{supp}(\chi)$, and set $\chi_n(\xi) = n^{1/(2\sigma)} \chi(n^{1/\sigma}\xi)$. With $a(\xi)$ as above, we have

$$\|\chi_n a^n\|_2^2 = \int_{-\infty}^{\infty} |\chi(\xi)|^2 |a(n^{-1/\sigma}\xi)|^{2n} d\xi ,$$

and since

$$\lim_{n\to\infty} |a(n^{-1/\sigma}\xi)|^n = \exp(-\gamma\xi^\sigma) ,$$

we find by dominated convergence that

$$\lim_{n\to\infty} \|\chi_n a^n\|_2^2 = \int |\chi(\xi)|^2 \exp(-2\gamma\xi^\sigma)d\xi > 0. \tag{3.7}$$

On the other hand, as in the proof of Corollary 1.5.1, Hölder's inequality and Parseval's formula give with $1/p + 1/p' = 1$,

$$\|\chi_n a^n\|_2^2 = 2\pi\|\mathcal{F}^{-1}(\chi_n a^n)\|_2^2$$

$$\le 2\pi\|\mathcal{F}^{-1}(a^n\chi_n)\|_{p'}\|\mathcal{F}^{-1}(\chi_n a^n)\|_2^{2/p}\|\mathcal{F}^{-1}(\chi_n a^n)\|_\infty^{1-2/p} \tag{3.8}$$

$$\le (2\pi)^{2\tilde{p}} M_p(e^n)\|\mathcal{F}^{-1}\chi_n\|_{p'}\|\chi_n a^n\|_2^{2/p}\|\mathcal{F}^{-1}(\chi_n a^n)\|_\infty^{2\tilde{p}} .$$

Since

$$(\mathcal{F}^{-1}\chi_n)(x) = n^{-1/(2\sigma)}(\mathcal{F}^{-1}\chi)(n^{-1/\sigma}x),$$

we obtain

$$\|\mathcal{F}^{-1}(\chi_n)\|_{p'} = n^{\tilde{p}/\sigma}\|\mathcal{F}^{-1}\chi\|_{p'} .$$

Hence (3.7) and (3.8) prove the existence of a positive constant c such that for large n,

$$(3.9) \qquad cn^{-\tilde{p}/\sigma} \leq M_p(e^n)\|\mathcal{F}^{-1}(\chi_n a^n)\|_\infty^{2\tilde{p}} .$$

We next estimate the maximum of $\mathcal{F}^{-1}(\chi_n a^n)$. With $x' = n^{-1/\sigma}x$, we have

$$(3.10) \qquad \mathcal{F}^{-1}(\chi_n a^n)(x) = (2\pi)^{-1}n^{-1/(2\sigma)} \int \exp(ix'\xi)\chi(\xi)a^n(n^{-1/\sigma}\xi)d\xi .$$

We consider $a^n(n^{-1/\sigma}\xi)$ in the support of χ, and write for small ξ, $a(\xi) = \exp(\rho(\xi))$, where for small ξ, by (3.4) and (3.5),

$$\rho(\xi) = i\beta\xi^\nu(1+o(1)), \ \mathrm{Re}\,\rho(\xi) = -\gamma\xi^\sigma(1+o(1)) .$$

Then

$$a^n(n^{-1/\sigma}\xi) = \exp(in^{1-\nu/\sigma}\rho_n(\xi))\omega_n(\xi) ,$$

where

$$\rho_n(\xi) = n^{\nu/\sigma}\,\mathrm{Im}\,\rho(n^{-1/\sigma}\xi), \ \omega_n(\xi) = \exp(n\,\mathrm{Re}\,\rho(n^{-1/\sigma}\xi)) .$$

For ξ in the support of χ we see that since $\nu \geq 2$, ρ_n'' is bounded away from zero, that ω_n' is bounded, and that both these bounds are uniform in n, for n large. Hence by van der Corput's lemma (Lemma 1.5.2) and (3.10),

$$(3.11) \qquad \|\mathcal{F}^{-1}(\chi_n a^n)\|_\infty \leq Cn^{-1/(2\sigma)}n^{-(1-\nu/\sigma)/2} = Cn^{-(1-(\nu-1)/\sigma)/2} .$$

In view of (3.9) this gives for some constant $c > 0$,

$$M_p(e^n) \geq cn^{\tilde{p}(1-\nu/\sigma)},$$

which is the desired lower bound. This completes the proof of the theorem.

For example, for the Lax-Wendroff operator $E_k = E_k^{(2)}$ defined in Section 2 we have $\nu = 3$ and $\sigma = 4$, so that

$$cn^{\tilde{p}/4} \leq \|\mathbb{E}_k^n\|_p \leq Cn^{\tilde{p}/4}.$$

As a corollary to the proof of Theorem 3.2 we note for later use the following:

<u>Corollary 3.1</u>. Let $g \in C_0^\infty$, $g \not\equiv 0$, and $a(\xi) = e(\xi)e^{-i\alpha\xi}$. Then under the assumptions of Theorem 3.2 there exists a constant $c > 0$ such that

$$M_p(ga^n(n^{-1/\sigma}\cdot)) \geq cn^{\tilde{p}(1-\nu/\sigma)}, \quad n = 1,2,\ldots .$$

<u>Proof</u>. Let χ be as in the proof of the estimate from below in Theorem 3.2, where now $\mathrm{supp}(\chi)$ is contained in the interior of $\mathrm{supp}(g)$. Let in addition $\tilde{\chi} \in C_0^\infty(\mathrm{supp}(g))$ with $\tilde{\chi} = 1$ on $\mathrm{supp}(\chi)$. Since $\tilde{\chi}\chi = \chi$, we may then replace e^n and a^n in (3.8) and (3.9) by $\tilde{\chi}(n^{1/\sigma}\xi)a^n(\xi)$. By (3.9), (3.10) and (3.11) we then obtain

$$M_p(\tilde{\chi}(n^{1/\sigma}\xi)a^n) \geq Cn^{\tilde{p}(1-\nu/\sigma)},$$

and since $\tilde{\chi}/g \in C_0^\infty \subseteq M_p$,

$$M_p(\tilde{\chi}(n^{1/\sigma}\xi)a^n) = M_p(\tilde{\chi}a(n^{-1/\sigma}\xi)^n) \leq M_p(\tilde{\chi}/g)M_p(ga(n^{-1/\sigma}\xi)^n),$$

which completes the proof of the corollary.

5.4. Convergence estimates.

We continue to consider the initial value problem (2.1) and a consistent difference operator E_k of the form (2.2). We shall now derive estimates for the rate of convergence in L_p for both stable and unstable E_k in terms of the smoothness of the data, measured in L_p. We shall also show that our results are in a sense best possible.

We shall first prove that for smooth initial data and L_2 stable difference operators we have convergence of order h^μ in L_p, where μ is the accuracy of the difference operator.

Theorem 4.1. Let $1 \leq p \leq \infty$, and assume that E_k is consistent with (2.1), accurate of order μ and stable in L_2. Then for each $T > 0$ there exists a constant C such that for $v \in B_p^{\mu+1,1}$,

$$\|E_k^n v - E(nk)v\|_p \leq Ch^\mu \|v\|_{B_p^{\mu+1,1}}, \quad \text{for } nk \leq T. \tag{4.1}$$

Proof. Setting

$$a(\xi) = e(\xi)e^{-i\lambda\xi}, \quad r_{nk}(\xi) = a(h\xi)^n - 1, \tag{4.2}$$

we may write

$$E_k^n v - E(nk)v = \mathcal{F}^{-1}((e(h\xi)^n - e^{ink\xi})\hat{v}) = \mathcal{F}^{-1}(e^{ink\xi} r_{nk}\hat{v}),$$

and by Lemma 2.6.2, (4.1) follows if

$$M_p(\phi_j r_{nk})2^{-j(\mu+1)} \leq Ch^\mu \quad \text{for } j \in Z. \tag{4.3}$$

By (4.2) and the L_2 stability we obtain for $\xi \in \mathrm{supp}(\phi_j)$,

$$|\phi_j(\xi)r_{nk}(\xi)| \leq Cn|a(h\xi) - 1| \leq Cn|h\xi|^{\mu+1} \leq Ch^\mu 2^{(\mu+1)j},$$

and since $a'(\xi) = O(\xi^\mu)$ for small ξ,

$$\left|\frac{d}{d\xi}(\phi_j(\xi)r_{nk}(\xi))\right| \leq C\{2^{-j}|r_{nk}(\xi)| + nh|a(h\xi)|^{n-1}|a'(h\xi)|\}$$

$$\leq C(h^\mu 2^{\mu j} + |h\xi|^\mu) \leq Ch^\mu 2^{\mu j}.$$

After integration this yields

$$(4.4) \qquad \left\|\left(\frac{d}{d\xi}\right)^l(\phi_j r_{nk})\right\|_2 \leq Ch^\mu 2^{j(\mu+3/2-l)}, \; l = 0,1,$$

and hence by the Carlson-Beurling inequality,

$$M_\infty(\phi_j r_{nk}) \leq Ch^\mu 2^{j(\mu+1)}.$$

This proves (4.3) and hence completes the proof of the theorem.

We shall now turn to convergence estimates for less smooth initial data. We consider first the case when the difference operator E_k is stable in L_p.

<u>Theorem 4.2</u>. Let $1 \leq p \leq \infty$, and assume that the operator E_k is consistent with (2.1), accurate of order μ and stable in L_p. Then for each $T > 0$ and s with $0 < s < \mu+1$, there exists a constant C such that for $v \in B_p^{s,\infty}$,

$$\|E_k^n v - E(nk)v\|_p \leq Ch^{\frac{s\mu}{\mu+1}} \|v\|_{B_p^{s,\infty}}, \text{ for } nk \leq T.$$

<u>Proof</u>. By the stability of E_k and the well-posedness of (2.1) we have

$$\|E_k^n v - E(nk)v\|_p \leq C\|v\|_p, \text{ for } nk \leq T.$$

Combining this with Theorem 4.1, the desired result follows by Corollary 2.5.1.

We shall now also treat the case of an operator E_k which is stable in L_2 but not necessarily stable in L_p for $p \neq 2$. Again we restrict the considerations to operators for which the characteristic function satisfies (3.1) or (3.2). In fact, we shall only carry out the analysis for (3.2), or more presicely, (3.4), (3.5) and (3.6). It is not difficult to see that the result below holds true for E_k satisfying (3.1) by formally setting $\sigma = \infty$. Recall that $\tilde{p} = |1/2 - 1/p|$.

Theorem 4.3. Let $1 \leq p \leq \infty$ and assume that E_k is consistent with (2.1), accurate of order μ and dissipative of order σ. Then for each $T > 0$ and s with $0 < s < \nu = \mu+1$, $s \neq \nu\tilde{p}$, there exists a constant C such that for $v \in B_p^{s,\infty}$,

$$\|E_k^n v - E(nk)v\|_p \leq Ch^{q(s)} \|v\|_{B_p^{s,\infty}}, \quad \text{for } nk \leq T, \tag{4.5}$$

where

$$q(s) = \begin{cases} s(1-1/\sigma) - \tilde{p}(1-\nu/\sigma) & \text{for } 0 < s < \nu\tilde{p}, \\ s\mu/\nu & \text{for } \nu\tilde{p} < s < \nu . \end{cases} \tag{4.6}$$

Proof. Since the case $\nu = \sigma$ is already covered by Theorem 4.2, we may assume here that $\nu < \sigma$. In order to prove (4.5) it is in fact sufficient to show that

$$\|E_k^n v - E(nk)v\|_p \leq Ch^{\tilde{p}\mu} \|v\|_{B_p^{\tilde{p}\nu,1}} . \tag{4.7}$$

For in view of Theorem 4.1, the desired result then follows for $\nu\tilde{p} < s < \nu$ by interpolation between (4.1) and (4.7) (Corollary 2.5.1). Similarly, since by Theorem 3.2 and the well-posedness of (2.1) in L_p,

$$\|E_k^n v - E(nk)v\|_p \leq Ch^{-\tilde{p}(1-\nu/\sigma)} \|v\|_p ,$$

interpolation between this and (4.7) proves the result for $0 < s < \nu\tilde{p}$.

In order to prove (4.7) we first notice that in the notation (4.2) of the proof of Theorem 4.1, it suffices to show that

$$M_p(\phi_j r_{nk}) 2^{-j\nu\tilde{p}} \leq Ch^{\mu\tilde{p}}, \quad \text{for } j \in Z . \tag{4.8}$$

We have at once

$$\|\phi_j r_{nk}\|_2 \leq 2\|\phi_j\|_2 \leq C2^{j/2} ,$$

and by (4.4),

$$\|\frac{d}{d\xi}(\phi_j r_{nk})\|_2 \le Ch^{\mu} 2^{j(\nu-1/2)} .$$

Hence by the Carlson-Beurling inequality,

$$M_\infty(\phi_j r_{nk}) \le Ch^{\mu/2} 2^{j\nu/2} .$$

This proves (4.8) for $p = \infty$. Since also obviously

$$M_2(\phi_j r_{nk}) \le C ,$$

the result follows for general p by Theorem 1.2.5. This completes the proof of the theorem.

As an example we exhibit in the following figure the graph of the convergence exponent $q(s)$ for the Lax-Wendroff operator $E_k^{(2)}$ $(\nu = 3, \sigma = 4)$ in the case $p = \infty$.

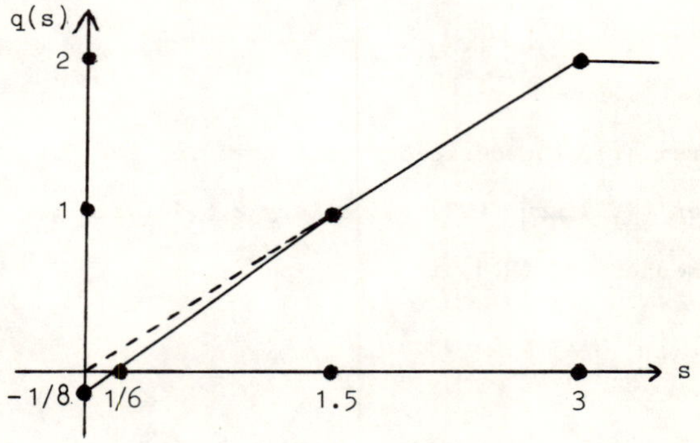

We shall now show that in a certain sense the exponent $q(s)$ in Theorem 4.3 is best possible. We shall say that E_k is accurate of order exactly μ if (3.4) holds with $\nu = \mu+1$ and $\beta \neq 0$.

Theorem 4.4. Let $1 \leq p \leq \infty$, $0 < s < \nu = \mu+1$ and let $t > 0$ be fixed. If E_k is consistent with (2.1), accurate of order exactly μ and dissipative of order σ, then there exist positive constants h_0 and c such that for $nk = t$ and $h \leq h_0$,

$$\sup\{\|\mathbb{E}_k^n v - E(nk)v\|_p : v \in B_p^{s,\infty},\ \|v\|_{B_p^{s,\infty}} \leq 1\} \geq ch^{q(s)},$$

where $q(s)$ is defined by (4.6).

Proof. By Theorem 2.2.2,

$$\|v\|_{B_p^{s,\infty}} \leq C\|v\|_{W_p^s},$$

and hence it is sufficient to show that for some positive c and h small,

$$m_h = \sup\{\|\mathbb{E}_k^n v - E(nk)v\|_p : v \in \hat{C}_0^\infty, \|v\|_{W_p^s} \leq 1\} \geq ch^{q(s)}.$$

Now with the notation of (4.2),

$$m_h = \sup\{\|\mathcal{F}^{-1}(r_{nk}\hat{v})\|_p : v \in \hat{C}_0^\infty, \|\mathcal{F}^{-1}((1+\xi^2)^{s/2}\hat{v})\|_p \leq 1\}$$

(4.9) $$= \sup\{\|\mathcal{F}^{-1}(r_{nk}(\xi)(1+\xi^2)^{-s/2}\hat{w})\|_p : w \in \hat{C}_0^\infty, \|w\|_p \leq 1\}$$

$$= M_p(r_{nk}(\xi)(1+\xi^2)^{-s/2}).$$

Since $\beta \neq 0$ in (3.4) we find for small $|h\xi|$, since $nk = t > 0$ is fixed,

(4.10) $$|r_{nk}(\xi)| \geq cn|h\xi|^\nu = ch^\mu|\xi|^\nu.$$

Now choose ξ_h such that $h^\mu\xi_h^\nu = 1$. Since $h\xi_h$ is small for h small, we then obtain by (4.9) and (4.10),

$$m_h \geq M_2(r_{nk}(\xi)(1+\xi^2)^{-s/2}) \geq |r_{nk}(\xi_h)(1+\xi_h^2)^{-s/2}| \geq ch^{s\mu/\nu}.$$

This proves the theorem in the stable case, and also for $\nu\tilde{p} < s < \nu$ in the unstable case.

It remains to consider the case $0 < s < \nu\tilde{p}$ in the unstable case. By Corollary 3.1, we have for $g \in C_0^\infty$, $g \not\equiv 0$, that

$$M_p(ga(n^{-1/\sigma}\xi)^n) \geq cn^{\tilde{p}(1-\nu/\sigma)} ,$$

and hence for large values of n,

$$M_p(g(a(n^{-1/\sigma}\xi)^n - 1)) \geq cn^{\tilde{p}(1-\nu/\sigma)} .$$

It follows, since $t = nk > 0$ is fixed, that for some $h_0 > 0$,

$$(4.11) \qquad M_p(gr_{nk}(h^{-(1-1/\sigma)}\xi)) \geq ch^{-\tilde{p}(1-\nu/\sigma)}, \ h \leq h_0 .$$

However, setting $h_\sigma = h^{1-1/\sigma}$ we obtain by the definition of m_h,

$$(4.12) \qquad M_p(gr_{nk}(h_\sigma^{-1}\xi)) = M_p(g(h_\sigma\xi)f_{nk}) \leq m_h M_p(g(h_\sigma\xi)(1+\xi^2)^{s/2}) .$$

Here we find easily for $h \leq h_0$,

$$\|(\tfrac{d}{d\xi})^l(g(h_\sigma\xi)(1+\xi^2)^{s/2})\|_2 \leq Ch_\sigma^{-s-1/2+l}, \ l = 0,1,$$

and hence the Carlson-Beurling inequality gives

$$M_p(g(h_\sigma\xi)(1+\xi^2)^{s/2}) \leq Ch_\sigma^{-s}, \quad h \leq h_0 .$$

From (4.11) and (4.12) we now conclude that

$$m_h \geq ch^{-\tilde{p}(1-\nu/\sigma)}h^{s(1-1/\sigma)} , \quad h \leq h_0$$

which completes the proof of the theorem.

We conclude by remarking that the estimate of Theorem 4.3 is not necessarily best possible for an individual function. Consider, for instance, the function H_s defined in Example II of Section 2.4 which is smooth except at the origin. For this function we have by Proposition 2.4.2, $H_s \in B_\infty^{s,\infty}$ and $H_s \in B_2^{s+1/2,\infty}$ but not

$H_s \in B_\infty^{s+\varepsilon,\infty}$ for any $\varepsilon > 0$. The convergence estimate in the maximum norm as determined by Theorem 3.3.8 is then $O(h^{s\mu/(\mu+1)})$ for small h when $0 < s < \mu+1$ which is an improvement over the result of Theorem 4.3 when $0 < s < \nu/2$.

5.5. Convergence estimates in a semi-linear problem.

In this section we shall consider the approximate solution of the model semi-linear hyperbolic equation

(5.1) $$\frac{\partial u}{\partial t} = \frac{\partial u}{\partial x} + u^2 \quad \text{for} \quad x \in R,\ t > 0,$$

with initial condition

(5.2) $$u(x,0) = v(x).$$

This problem has the exact solution

(5.3) $$u(x,t) = S(t)v = \frac{v(x+t)}{1 - tv(x+t)} \quad \text{for} \quad t\|v\|_\infty < 1.$$

Replacing derivatives in (5.1) by forward finite difference quotients we obtain a finite difference scheme for (5.1), (5.2) which consists in approximating $u(\cdot,nk)$ by $S_k^n v$ where

(5.4) $$S_k v = E_k v + kF_k v,$$

and where with $\lambda = k/h = \text{constant}$,

(5.5) $$E_k v(x) = \lambda v(x+h) + (1-\lambda)v(x),\ F_k v(x) = v(x)^2 .$$

The linear operator E_k (the operator $E_k^{(1)}$ of Section 2) is stable in the maximum-norm for $\lambda \leq 1$, and the non-linear operator F_k has the property that

(5.6) $$\|F_k v\|_\infty = \|v^2\|_\infty = \|v\|_\infty^2 .$$

Using these facts, it is possible to prove convergence estimates for this scheme. However, since the linear operator E_k is only first order accurate, the convergence will only be first order even for smooth initial data.

Assume now that we want to apply instead a second order accurate operator S_k of the form (5.4), based for instance on choosing for E_k the Lax-Wendroff operator (the operator $E_k^{(2)}$ of Section 2),

$$(5.7) \qquad E_k v(x) = \tfrac{1}{2}(\lambda^2+\lambda)v(x+h) + (1-\lambda^2)v(x) + \tfrac{1}{2}(\lambda^2-\lambda)v(x-h),$$

and some suitable non-linear operator F_k. For a smooth solution of (5.1) we have with $u_x = \partial u/\partial x$ and $u_t = \partial u/\partial t$,

$$\begin{aligned} u(x,t+k) &= u + ku_t + \tfrac{1}{2}k^2u_{tt} + o(k^2) \\ (5.8) \qquad &= [u+ku_x + \tfrac{1}{2}k^2u_{xx}] + k[u^2 + k(u^2)_x + ku^3] + o(k^2) \\ &= E_k u(x,t) + k[u^2 + k(u^2)_x + ku^3] + o(k^2), \text{ as } k \to 0, \end{aligned}$$

and we could therefore take

$$(5.9) \qquad F_k v(x) = v(x)^2 + \tfrac{1}{2}\lambda v(x+h)^2 - \tfrac{1}{2}\lambda v(x-h)^2 + kv(x)^3 .$$

To be concrete we shall consider from now on only the scheme defined by (5.7) and (5.9).

The operator (5.7) is accurate of order exactly 2, and although stable in L_2 for $\lambda \le 1$, it is unstable in the maximum-norm for $\lambda < 1$ by Theorem 2.1. On the other hand, the analogue of (5.6) does not hold in L_2, and therefore L_2 is also not suited for the analysis. It will be seen that a convenient basic space of functions to work with here is the Besov space $B_2^{1/2,1}$, which for brevity most often will be denoted by B below. The reason for this choice is that the operator E_k retains in B its L_2 stability property, and that by Theorem 2.2.4 the norm in B, denoted by $\|\cdot\|$, majorizes the maximum norm,

$$(5.10) \qquad \|v\|_\infty \le \mu_0 \|v\| .$$

As a consequence of the latter inequality we shall see that the norm in B is sub-multiplicative,

$$\|vw\| \leq \mu\|v\| \cdot \|w\| ,$$

which will replace the corresponding property (5.6) for the scheme (5.4), (5.5).

Our main convergence result is that as h tends to zero we have

$$(5.11) \qquad \|u(\cdot,nk) - S_k^n v\| = \begin{cases} O(h^2) & \text{if } v \in B_2^{7/2,1} \\ O(h^{2s/3}), & \text{if } v \in B_2^{1/2+s,\infty}, \; 0 < s < 3 . \end{cases}$$

These error estimates will hold for $t = nk$ in the whole life span of the solution (5.3), uniformly on compact subsets. They generalize to the present non-linear situation the results of Theorems 3.3.7 and 3.3.8 (rather than those of Section 4).

We first outline the main steps of the proof of (5.11). Introducing the error at time jk, $w_j = S_k^j v - S(jk)v$, we may write

$$w_{j+1} = E_k w_j + k[F_k S_k^j v - F_k S(jk)v] + [(S_k - S(k))S(jk)v] .$$

Estimating the three terms on the right we shall obtain for $v \in B_2^{7/2,1}$, using a stability property of E_k in B, the definition of F_k, and a calculation similar to (5.8), respectively, an estimate of the form

$$\|w_{j+1}\| \leq (1+\gamma_1 k)\|w_j\| + \gamma_2 kh^2 ,$$

where $\gamma_1 = \gamma_1(\max(\|S_k^j v\|, \|S(jk)v\|))$, $\gamma_2 = \gamma_2(S(jk)v)$. If the coefficients γ_1 and γ_2 were uniformly bounded in j, the first result of (5.11) would follow immediately by iteration, since $w_0 = 0$. In order to secure this boundedness we have to step forward in j in such a way as to keep control over $S_k^j v$ and $S(jk)v$.

In the case of less smooth v we shall express the error with the help of an approximating smooth w as

$$S_k^n v - S(nk)v = [S_k^n w - S(nk)w] + [S_k^n v - S_k^n w - (S(nk)v - S(nk)w)]$$

from which we shall be able to derive, by means of the result already obtained for w,

$$\|S_k^n v - S(nk)v\| \leq \gamma_3 h^2 + \gamma_4 \|v-w\| ,$$

were again γ_3 depends on the smoothness of w and γ_4 depends on bounds for $S_k^j v$, $S_k^j w$ for $j \leq n$. The desired result will now follow by choosing w in an optimal fashion, and again taking precautions to control the size of the constants.

We shall now begin the technical work. Rather than the norm in $B_2^{1/2,1}$ described in Section 2.1, we shall employ the equivalent norm described in Section 2.3, viz.

$$\|v\| = \|v\|_2 + \dot{B}_2^{1/2,1}(v) = \|v\|_2 + \int_0^\infty \frac{\omega_2(v;s)}{s^{3/2}}\, ds ,$$

where ω_2 is the modulus of continuity in L_2. This norm is sub-multiplicative:

Lemma 5.1. There exists a constant μ such that if $v, w \in B$ then $vw \in B$ and

$$\|vw\| \leq \mu \|v\|\, \|w\| .$$

Proof. We have by (5.10),

(5.12) $$\|vw\|_2 \leq \|v\|_\infty \|w\|_2 \leq \mu_0 \|v\|\, \|w\| .$$

We also find

(5.13) $$\omega_2(vw;s) \leq \|v\|_\infty \omega_2(w;s) + \|w\|_\infty \omega_2(v;s) ,$$

so that

$$\int_0^\infty \frac{\omega_2(vw;s)}{s^{3/2}}\, ds \leq \|v\|_\infty \int_0^\infty \frac{\omega_2(w;s)}{s^{3/2}}\, ds + \|w\|_\infty \int_0^\infty \frac{\omega_2(v;s)}{s^{3/2}}\, ds .$$

In view of (5.12) this proves the lemma with $\mu = 3\mu_0$.

Together with $B = B_2^{1/2,1}$ we shall also employ below, for s positive, the space $B_2^{s+1/2,\infty}$, which will be denoted by B^s, with norm

$$\|v\|_{B^s} = \|v\|_{B_2^{s+1/2,\infty}} .$$

In addition, for s a positive integer, we shall work with $B_2^{s+1/2,1}$, which will be denoted by $B^{(s)}$. Recall that $B^{(s)}$ is the space of $v \in B$ for which also the derivatives $v^{(j)} \in B$ for $j \leq s$.

For $v \in B^{(s)}$, $s = 1,2,3$, we introduce the positive functionals

$$N_1(v) = \|v'\|, \quad N_2(v) = \|v''\| + \|v'\|^2, \quad N_3(v) = \|v'''\| + \|v''\|\,\|v'\| + \|v'\|^3 .$$

The following non-linear interpolation lemma will be useful in the treatment of non-smooth initial data.

Lemma 5.2. Let $0 < s < 3$, and let $v \in B^s$. Then there exists a constant $C_s(v)$ and for each ε with $0 < \varepsilon < 1$ a $v_\varepsilon \in B^{(3)}$ such that

$$\varepsilon^{(3-s)} N_3(v_\varepsilon) + \varepsilon^{-s}\|v-v_\varepsilon\| + \|v_\varepsilon\| \leq C_s(v) .$$

Proof. Set in the notation of Section 2.1

$$v_\varepsilon = \mathcal{F}^{-1}\Big(\sum_{2^j\varepsilon \leq 1} \psi_j \hat{v}\Big), \quad v-v_\varepsilon = \mathcal{F}^{-1}\Big(\sum_{2^j\varepsilon > 1} \psi_j \hat{v}\Big) .$$

We obtain, since $\psi_l \psi_j = 0$ for $|j-l| > 1$,

$$\begin{aligned} \|v-v_\varepsilon\| &\leq C \sum_{l=0}^{\infty} 2^{l/2} \Big\|\psi_l \sum_{2^j\varepsilon > 1} \psi_j \hat{v}\Big\|_2 \leq C \sum_{2^{l+1}\varepsilon > 1} 2^{l/2} \|\psi_l \hat{v}\|_2 \\ &\leq C\Big(\sum_{2^{l+1}\varepsilon > 1} 2^{-sl}\Big)\|v\|_{B_2^{s+1/2,\infty}} \leq k_1 \varepsilon^s \|v\|_{B^s} . \end{aligned} \tag{5.14}$$

(In this proof k_j denote constants depending only on s.) In particular,

$$\|v_\varepsilon\| \leq \|v\| + k_1\|v\|_{B^s} . \tag{5.15}$$

Similarly, since $\operatorname{supp}(\psi_l) \subset \{\xi : 2^{l-1} < |\xi| < 2^{l+1}\}$, for $l > 0$,

$$\|v'''\| \leq C \sum_{l=0}^{\infty} 2^{l/2} \Big\|\psi_l \sum_{2^j\varepsilon \leq 1} \psi_j \xi^3 \hat{v}\Big\|_2 \leq k_2 \sum_{2^{l-1}\varepsilon \leq 1} 2^{(3-s)l} \|v\|_{B^s} \leq k_3 \varepsilon^{-(3-s)} \|v\|_{B^s} .$$

We also have

$$\|v_\varepsilon''\| \le k_4 \sum_{\substack{2^{l-1}\varepsilon \le 1\\ l>0}} 2^{(2-s)l}\|v\|_{B^s} \le k_5\|v\|_{B^s} \begin{cases} \varepsilon^{-(2-s)}, & \text{for } s<2,\\ \log 1/\varepsilon, & \text{for } s=2,\\ 1, & \text{for } s>2,\end{cases}$$

and

$$\|v_\varepsilon'\| \le k_6\|v\|_{B^s} \begin{cases} \varepsilon^{-(1-s)} & \text{for } s<1,\\ \log 1/\varepsilon & \text{for } s=1,\\ 1 & \text{for } s>1.\end{cases}$$

For all s with $0 < s < 3$ we find

$$N_3(v_\varepsilon) \le k_7\varepsilon^{-(3-s)}(\|v\|_{B^s} + \|v\|^2_{B^s} + \|v\|^3_{B^s}),$$

which together with (5.14) and (5.15) proves the lemma.

We shall next collect some simple estimates for the solution (5.3) of the initial value problem (5.1), (5.2). For bounded, continuous v we define the (generalized) solution operator $S(t)$ of this problem by

$$\text{(5.16)}\qquad S(t)v(x) = \frac{v(x+t)}{1-tv(x+t)}, \quad \text{for } 0 \le t < \|v\|_\infty^{-1}.$$

Throughout the rest of this section we shall let T and T_0 be positive constants with $0 < T < T_0$. If we then let v vary subject to the condition $\|v\|_\infty T_0 < 1$, then for $0 \le t \le T$ the denominator in (5.16) is bounded away from zero; in fact

$$\text{(5.17)}\qquad \|(1-tv)^{-1}\|_\infty \le \kappa = (1-TT_0^{-1})^{-1}.$$

The following lemma expresses in quantitative form the fact that if the initial function is in B then so is the solution of (5.1), (5.2).

Lemma 5.3. There exists a positive constant β_0 such that for all $v \in B$ with $\|v\|_\infty T_0 < 1$ we have $S(t)v \in B$ and

$$\|S(t)v\| \le \beta_0\|v\|, \text{ for } 0 \le t \le T.$$

Proof. We find at once by (5.17),

$$\|S(t)v\|_2 \leq \kappa \|v\|_2 . \tag{5.18}$$

Further, since for $u(\cdot,t) = S(t)v$,

$$u(x+s,t) - u(x,t) = \frac{v(x+s+t) - v(x+t)}{(1-tv(x+t))(1-tv(x+s+t))},$$

we obtain

$$\omega_2(S(t)v;s) \leq \kappa^2\omega_2(v;s) .$$

Together with (5.18) this proves the lemma with $\beta_0 = \kappa^2$.

We shall now see that the solution $S(t)v$ is smooth with v and derive certain bounds for its derivatives.

Lemma 5.4. There exists an increasing function β such that if $v \in B^{(3)}$ and $\|v\|_\infty T_0 < 1$, then we have $S(t)v \in B^{(3)}$ for $0 \leq t \leq T$ and

$$\|D^j S(t)v\| \leq \beta(\|v\|)N_j(v) \quad \text{for } j = 1,2,3, \tag{5.19}$$

$$\left\|\frac{\partial^3}{\partial t^3} S(t)v\right\| \leq \beta(\|v\|)(N_3(v)+1) . \tag{5.20}$$

Proof. Consider generally for $0 \leq t \leq T$ a function of the form $w(x,t) = w_1(x)(1-tw_2(x))^{-1}$ where $\|w_2\|_\infty T_0 < 1$. We have at once

$$\|w(\cdot,t)\|_2 \leq \kappa\|w_1\|_2 .$$

Further, we find by (5.17) and (5.10),

$$\omega_2(w(\cdot,t);s) \leq \|w_1\|_\infty \omega_2((1-tw_2)^{-1};s) + \|(1-tw_2)^{-1}\|_\infty \omega_2(v;s)$$

$$\leq \mu_0\|w_1\|\kappa^2 T\omega_2(w_2;s) + \kappa\omega_2(w_1;s) ,$$

and hence

$$\|w(\cdot,t)\| \leq (\kappa + \mu_0\kappa^2 T\|w_2\|)\|w_1\| .$$

Noticing now that

$$(5.21) \qquad Dw(\cdot,t) = \frac{w_1'}{1-tw_2} + \frac{w_1}{1-tw_2} \cdot \frac{w_2'}{1-tw_2} ,$$

we find in particular that $DS(t)v \in B$ and that the inequality (5.19) holds for $j = 1$. After differentiation once and twice of (5.21) the same argument proves that $D^jS(t)v \in B$ and that (5.19) holds for $j = 2,3$. Using the differential equation (5.1) we find for $u(\cdot,t) = S(t)v$,

$$\frac{\partial^3}{\partial t^3} u(\cdot,t) = D^3u + 6uD^2u + 6(Du)^2 + 18u^2Du + 6u^4 ,$$

from which we see that (5.20) follows from Lemma 5.1 and (5.19).

We shall now study the effect of changing the initial data in (5.1), (5.2).

Lemma 5.5. There exists an increasing function $\tilde{\beta}$ such that for all $v,w \in B$ with $\max(\|v\|_\infty,\|w\|_\infty)T_0 < 1$ we have

$$\|S(t)v - S(t)w\| \leq \tilde{\beta}(\max(\|v\|,\|w\|))\|v-w\|, \text{ for } 0 \leq t \leq T .$$

Proof. We have

$$S(t)v - S(t)w = \frac{v(x+t) - w(x+t)}{(1-tv(x+t))(1-tw(x+t))} ,$$

so that

$$(5.22) \qquad \|S(t)v - S(t)w\|_2 \leq \kappa^2\|v-w\|_2 .$$

Further, by (5.13),

$$\omega_2(S(t)v - S(t)w;s)$$

$$\leq \|(1-tv)^{-1}(1-tw)^{-1}\|_\infty \omega_2(v-w;s) + \|v-w\|_\infty \omega_2((1-tv)^{-1}(1-tw)^{-1};s) .$$

Here

$$\omega_2((1-tv)^{-1}(1-tw)^{-1};s)$$

$$\leq \|(1-tv)^{-1}\|_\infty \omega_2((1-tw)^{-1};s) + \|(1-tw)^{-1}\|_\infty \omega_2((1-tv)^{-1};s)$$

$$\leq \kappa^3 T(\omega_2(w;s) + \omega_2(v;s)),$$

and hence we obtain in the same way as above,

$$\omega_2(S(t)v - S(t)w;s) \leq \kappa^2 \omega_2(v-w;s) + \mu_0\|v-w\|\kappa^3 T(\omega_2(w;s) + \omega_2(v;s)).$$

Together with (5.22) this yields

$$\|S(t)v - S(t)w\| \leq (\kappa^2 + \mu_0\kappa^3 T(\|v\| + \|w\|))\|v-w\|,$$

which proves the lemma with $\tilde{\beta}(y) = \kappa^2 + 2\mu_0\kappa^3 Ty$.

We shall now prove some stability estimates for the discrete solution operator $S_k = E_k + kF_k$ with E_k and F_k defined by (5.7) and (5.9). Our first lemma concerns the stability in B of the linear operator E_k.

Lemma 5.6. We have for $\lambda \leq 1$,

$$\|E_k v\| \leq \|v\|, \quad \text{for } v \in B.$$

Proof. As we already know from Section 2, for these λ,

$$\|E_k v\|_2 \leq \|v\|_2.$$

From this follows immediately,

$$\omega_2(E_k v;s) = \sup_{|\eta|\leq s} \|E_k v(\cdot+\eta) - E_k v\|_2 \leq \omega_2(v;s),$$

and hence the lemma follows by the definition of the norm in B.

We next deduce some estimates for the non-linear part F_k of S_k.

Lemma 5.7. For $\lambda \leq 1$, $k \leq 1$ we have

$$\|F_k v\| \leq f(\|v\|) \quad \text{for} \quad v \in B,$$

$$\|F_k v - F_k w\| \leq f'(\max(\|v\|,\|w\|))\|v-w\| \quad \text{for} \quad v,w \in B,$$

where $f(y) = 2\mu y^2 + \mu^2 y^3$, $f'(y) = 4\mu y + 3\mu^2 y^2$.

Proof. We have from the definition of F_k, using Lemma 5.1,

$$\|F_k v\| \leq (1+\lambda)\|v^2\| + k\|v^3\| \leq 2\mu\|v\|^2 + \mu^2\|v\|^3 = f(\|v\|),$$

and also, with $v_{\pm}(x) = v(x\pm h)$,

$$\|F_k v - F_k w\|$$

$$= \|(v-w)((v+w) + k(v^2+vw+w^2)) + \frac{1}{2}\lambda(v_+-w_+)(v_++w_+) - \frac{1}{2}\lambda(v_--w_-)(v_-+w_-)\|$$

$$\leq \|v-w\|[(1+\lambda)\mu(\|v\| + \|w\|) + k\mu^2(\|v\|^2 + \|v\|\ \|w\| + \|w\|^2)]$$

$$\leq f'(\max(\|v\|,\|w\|))\|v-w\|.$$

This proves the lemma.

We shall now prove the boundedness of the iterates S_k^n for nk in a small interval.

Lemma 5.8. For γ a given positive constant there exists a positive constant τ such that for all $v \in B$ with $\|v\| \leq 3\gamma$ we have $\|S_k^n v\| \leq 4\gamma$, for $nk < \tau$.

Proof. Let $f(y)$ be the function of Lemma 5.7, let $y(t)$ be the solution of the initial value problem

$$y' = f(y), \quad y(0) = 3\gamma,$$

and choose τ so small that $y(\tau) \leq 4\gamma$. Setting $\sigma_n = \max_{j \leq n} \|S_k^j v\|$, we shall prove by induction over n that $\sigma_n \leq y(nk)$ for $nk \leq \tau$. Thus assume that $\sigma_n \leq y(nk)$. We obtain directly from Lemmas 5.6 and 5.7,

$$\|S_k^{n+1}v\| \leq \|E_k S_k^n v\| + k\|F_k S_k^n v\| \leq \|S_k^n v\| + kf(\|S_k^n v\|),$$

and hence, using the induction assumption and our definitions,

$$\sigma_{n+1} \leq \sigma_n + kf(\sigma_n) \leq y(nk) + \int_{nk}^{(n+1)k} f(y(s))ds = y((n+1)k)\,.$$

Since $\sigma_0 = \|v\| \leq 3\gamma = y(0)$, this completes the proof.

The last lemma before we begin the convergence estimates is a continuity estimate for the discrete solution.

Lemma 5.9. For γ a given positive constant there exists a constant $\tilde{b} > 1$ such that

$$\|S_k^n v - S_k^n w\| \leq \tilde{b}\|v-w\| \quad \text{for} \quad nk \leq T,$$

so long as $\max(\|S_k^j v\|, \|S_k^j w\|) \leq 4\gamma$ for $j < n$.

Proof. Again we use induction over n and notice that the inequality trivially holds for $n = 0$ if $\tilde{b} \geq 1$. We have for the step from n to n+1,

$$S_k^{n+1}v - S_k^{n+1}w = E_k(S_k^n v - S_k^n w) + k(F_k S_k^n v - F_k S_k^n w)\,,$$

and hence by Lemmas 5.6 and 5.7,

$$\|S_k^{n+1}v - S_k^{n+1}w\| \leq (1 + kf'(4\gamma))\|S_k^n v - S_k^n w\|\,.$$

The result now follows with $\tilde{b} = \exp(Tf'(4\gamma))$.

We now turn to the estimates for the difference between the exact and approximate solutions of the initial value problem (5.1), (5.2), first for smooth and then for less smooth initial values. For this purpose we shall investigate the truncation error for smooth data and define the local truncation error,

$$\varepsilon_k(t,v) = \|S_k u(t) - S(k)u(t)\| ,$$

where $u(t) = u(\cdot,t)$ is the exact solution of (5.1), (5.2). We also define the global truncation error,

$$(5.23) \qquad \tilde{\varepsilon}_k(v) = \sum_{j=0}^{[T/k]} \varepsilon_k(jk,v) .$$

We shall prove the following:

Lemma 5.10. There exists an increasing function ρ such that for $v \in B^{(3)}$ and $\|v\|_\infty T_0 < 1$, we have uniformly for $0 \leq t \leq T$,

$$(5.24) \qquad \varepsilon_k(t,v) \leq kh^2\rho(\|v\|)(N_3(v) + 1) .$$

Further,

$$\tilde{\varepsilon}_k(v) \leq h^2 T\rho(\|v\|)(N_3(v) + 1).$$

Proof. With $u_t = \partial u/\partial t$, $u_x = \partial u/\partial x$ and supressing the arguments at (x,t), we have by Taylor's formula and the differential equation (5.1),

$$(5.25) \qquad \begin{aligned} (S(k)u(t))(x) &= u(x,t+k) = u + ku_t + \tfrac{1}{2}k^2u_{tt} + R_t \\ &= [u + ku_x + \tfrac{1}{2}k^2u_{xx}] + k[u^2 + k(u^2)_x + ku^3] + R_t , \end{aligned}$$

where

$$\|R_t\| = \|\tfrac{1}{2}\int_0^k (s-k)^2 u_{ttt}(\cdot,t+s)ds\| \leq \tfrac{1}{2}\int_0^k (s-k)^2\|u_{ttt}(\cdot,t+s)\| ds .$$

Using Lemma 5.4 we obtain

$$(5.26) \qquad \|R_t\| \leq \tfrac{1}{6}k^3\beta(\|v\|)(N_3(v) + 1),$$

which is of the form of the right hand side of (5.24).

By Taylor's formula we also obtain

$$E_k w = w+kw' + \frac{1}{2}k^2 w'' + \frac{1}{4}(\lambda+\lambda^2) \int_0^h (s-h)^2 w'''(x+s)ds - \frac{1}{4}(\lambda-\lambda^2) \int_{-h}^0 (s+h)^2 w'''(x+s)ds$$

and hence, since the norm in B is translation invariant and $\lambda \leq 1$,

$$\|E_k w - w - kw' - \frac{1}{2}k^2 w''\|$$

$$\leq \{\frac{1}{4}(\lambda+\lambda^2) \int_0^h (s-h)^2 ds + \frac{1}{4}(\lambda-\lambda^2) \int_{-h}^0 (s+h)^2 ds\} \|w'''\| = \frac{1}{6}kh^2 \|w'''\| .$$

With $w = u(\cdot,t)$ and using Lemma 5.4 this yields

$$(5.27) \qquad \|E_k u(t) - u - ku_x - \frac{1}{2}k^2 u_{xx}\| \leq \frac{1}{6}kh^2 \beta(\|v\|) N_3(v) .$$

Similarly we obtain

$$\|kF_k w - k[w^2+k(w^2)' + kw^3]\| = k^2 \| \frac{w(x+h)^2 - w(x-h)^2}{2h} - (w^2)'(x)\|$$

$$= k\lambda \| \int_0^h (h-s)[(w^2)''(x+s) - (w^2)''(x-s)]ds \|$$

$$\leq \lambda kh^2 \|(w^2)''\| \leq 2\lambda kh^2 (\|w''\| \, \|w\| + \|w'\|^2) .$$

Applying this with $w = u(\cdot,t)$ and using Lemma 5.4 we obtain a bound of the form of the right hand side of (5.24). Together with (5.25), (5.26) and (5.27) this proves the estimate for the local truncation error.

The estimate for the global truncation error now follows immediately by summation.

In the error analysis for smooth v, we shall first prove an estimate under a boundedness assumption for the discrete solution.

Lemma 5.11. Let $v \in B^{(3)}$, $\|v\|_\infty T_0 < 1$, and assume that $\|S_k^j v\| \leq 4\gamma$ for $j < n$ where $\gamma = \beta_0\|v\|$ and β_0 is the constant of Lemma 5.3. Then

$$\|S_k^n v - S(nk)v\| \leq C_0 \tilde{\varepsilon}_k(v) \quad \text{with} \quad C_0 = \exp(T_0 f'(4\gamma)),$$

where $\tilde{\varepsilon}_k(v)$ is the global truncation error (5.23) and f' is the function of Lemma 5.7.

Proof. Setting

$$w_j = S_k^j v - S(jk)v = S_k^j v - u(jk)$$

we have for $j < n$,

$$w_{j+1} = S_k S_k^j v - u((j+1)k)$$

$$= E_k w_j + k[F_k S_k^j v - F_k(S(jk)v)] + S_k u(jk) - u((j+1)k),$$

and hence, by Lemmas 5.6 and 5.7,

$$\|w_{j+1}\| \leq (1 + kf'(4\gamma))\|w_j\| + \varepsilon_k(jk,v).$$

Since $w_0 = 0$ it follows that

$$\|w_n\| \leq \sum_{j=0}^{n-1} (1 + kf'(4\gamma))^{n-1-j} \varepsilon_k(jk,v) \leq \exp(T_0 f'(4\gamma))\tilde{\varepsilon}_k(v),$$

which proves the lemma.

We can now prove the following convergence estimate.

Theorem 5.1. Assume that $v \in B^{(3)} = B_2^{7/2,1}$, $\|v\|_\infty T_0 < 1$ and $T < T_0$. Then there exist a $c_1 = c_1(v)$ and a positive $h_1 = h_1(v)$ such that for $nk \leq T$, $h \leq h_1$ we have

$$\|S_k^n v - S(nk)v\|_{B_2^{1/2,1}} \leq c_1 h^2.$$

Proof. Let again $\gamma = \beta_0\|v\|$ where β_0 is the constant of Lemma 5.3. We shall prove that if h_1 is chosen so that $c_0\tilde{\varepsilon}_k(v) < \gamma$ for $h \le h_1$, where c_0 is the constant of Lemma 5.11, then for these h we have $\|S_k^n v\| \le 4\gamma$ for $nk \le T$. In view of Lemmas 5.11 and 5.10 this proves the theorem.

Let now τ be the number of Lemma 5.8 and set $M = [\tau k^{-1}]$ and $t_j = Mjk$, so that $0 = t_0 < t_1 < \dots$ with $t_{j+1} - t_j < \tau$. Assume that we have already proved $\|S_k^n v\| \le 4\gamma$ for $nk \le t_j$, and let $n = Mj + l$, with $l \le M$. We have by the choice of h_1 and by Lemmas 5.3 and 5.11,

$$\|S_k^{Mj} v\| \le \|S(tj)v\| + \|S_k^{Mj} v - S(Mjk)v\| \le 2\gamma .$$

Hence by Lemma 5.8,

$$\|S_k^n v\| = \|S_k^l S_k^{Mj} v\| \le 4\gamma ,$$

which completes the proof of the theorem.

We shall now treat the case when the initial data belong to a space B^s with $0 < s < 3$. This will be done by an interpolation argument between B and $B^{(3)}$. We shall then need the following continuity property of the discrete solution operator:

Lemma 5.12. Let $v \in B$ be given and let γ be as in Lemma 5.11. Then there exists a positive constant δ such that if $\|S_k^l v\| \le 2\gamma$ for $l \le n$ and $\|w-v\| \le \delta$ then $\|S_k^l w\| \le 3\gamma$ for $l \le n$.

Proof. Let the numbers $t_j = Mjk$ be as in the proof of Theorem 5.1, and assume that the result holds for $nk \le t_j$. It follows then by Lemma 5.8 that for $n = Mj+l$, with $l \le M$, we have

$$\|S_k^n v\| \le 4\gamma, \quad \|S_k^n w\| \le 4\gamma ,$$

and hence by Lemma 5.9,

$$\|S_k^n v - S_k^n w\| \le \tilde{b}\|v-w\| \le \gamma \quad \text{if} \quad \|w-v\| \le \delta = \gamma/\tilde{b} ,$$

so that for these w,

$$\|S_k^n w\| \leq \|S_k^n v\| + \gamma \leq 3\gamma .$$

Since for $j = 0$, we have $\|w\| \leq \|v\| + \|w-v\| \leq 3\gamma$ if $\|w-v\| \leq \delta < \gamma$, this completes the proof.

We can now finally prove the following convergence result.

<u>Theorem 5.2</u>. Assume that $0 < s < 3$, $v \in B^s = B_2^{s+1/2,\infty}$, $\|v\|_\infty T_0 < 1$ and $T < T_0$. Then there exist a $c_2 = c_2(v)$ and a positive $h_2 = h_2(v)$ such that for $nk \leq T$, $h \leq h_2$, we have

$$\|S_k^n v - S(nk)v\|_{B_2^{1/2,1}} \leq c_2 h^{2s/3} .$$

<u>Proof</u>. Again with the notation of the proof of Theorem 5.1, assume that the result holds for $nk \leq t_j$ with some choice of c_2 and h_2. In particular, we then have for $nk \leq t_j$, $h \leq h_2$,

$$\|S_k^n v\| \leq \|S(nk)v\| + c_2 h^{2s/3} \leq 2\gamma \quad \text{if} \quad c_2 h^{2s/3} \leq \gamma \quad \text{for} \quad h \leq h_2 .$$

Hence, by Lemma 5.12, if $\|w-v\| \leq \delta$ we have $\|S_k^n w\| \leq 3\gamma$ for $nk \leq t_j$, and it follows by Lemma 5.8, for these w,

$$\max(\|S_k^n v\|, \|S_k^n w\|) \leq 4\gamma, \text{ for } nk \leq t_{j+1} .$$

By Lemma 5.9 we hence have for $\|w-v\| \leq \delta$, $nk \leq t_{j+1}$,

$$\|S_k^n v - S_k^n w\| \leq \tilde{b}\|v-w\| .$$

Also, by Lemma 5.5, if $\|w-v\| \leq \delta$ we have for $nk \leq t_{j+1}$,

$$\|S(nk)v - S(nk)w\| \leq \tilde{\beta}\|v-w\|, \quad \text{with} \quad \tilde{\beta} = \tilde{\beta}(\gamma+\delta).$$

Let now $\delta_1 \leq \delta$ be so small that for $\|w-v\| \leq \delta_1$ in addition $\|w\|_\infty T_0 < 1$, which is possible since by (5.10),

$$\|w\|_\infty T_0 \leq \|v\|_\infty T_0 + \mu_0 \|v-w\| T_0 \leq \|v\|_\infty T_0 + \mu_0 \delta_1 T_0 .$$

By Lemma 5.11 we then have for $w \in B^{(3)}$ with $\|w-v\| \leq \delta_1$ and $nk \leq t_{j+1}$,

$$\|S_k^n w - S(nk)w\| \leq c_0 \tilde{\varepsilon}_k(w) .$$

For such w and $nk \leq t_{j+1}$, we may therefore write, using Lemma 5.10 in the last step,

$$\|S_k^n v - S(nk)v\|$$

$$\leq \|S_k^n w - S(nk)w\| + \|S_k^n v - S_k^n w\| + \|S(nk)v - S(nk)w\|$$

$$\leq c_0 \tilde{\varepsilon}_k(w) + (\tilde{b}+\tilde{\beta})\|v-w\| \leq h^2 T\rho(\|w\|)(N_3(w)+1) + (\tilde{b}+\tilde{\beta})\|v-w\| .$$

We now choose $w = v_\varepsilon$ as in Lemma 5.2 with $\varepsilon > 0$ so small that $\|v-v_\varepsilon\| \leq \delta_1$. We then have

$$\|S_k^n v - S(nk)v\| \leq h^2 T\rho(C_s(v))(C_s(v)\varepsilon^{-(3-s)} + 1) + (\tilde{b}+\tilde{\beta})\varepsilon^s C_s(v) .$$

Taking $\varepsilon = h^{2/3}$ with $h \leq 1$, this proves the result with $c_2 = T\rho(C_s(v))(C_s(v)+1) + (\tilde{b}+\tilde{\beta})C_s(v)$ and with h_2 such that all the above conditions are satisfied, i.e. such that for $h \leq h_2$ we have $c_2 h^{2s/3} \leq \gamma$, $h \leq 1$ and $\|v-v_\varepsilon\| \leq \delta_1$. This completes the proof of the theorem.

<u>References</u>.

The result in Section 1 was proved in [2] and further developed in [3]. The stability and convergence theory in Sections 2 through 4 is reproduced from [4]; see also [5], [6], [7], [8], [9] and references for related material. For a convergence study in the maximum norm in the case of an ill-posed (with respect to this norm) hyperbolic system, see [11]. The analysis in Section 5 was presented in [10]; for the operator defined in (5.5) the corresponding investigation was carried out in [1].

1. R. Ansorge, C. Geiger and R. Hass, Existenz und numerische Erfassbarkeit verallgemeinerter Lösungen halblinearer Anfangswertaufgaben, Z. Angew. Math. Mech. 52 (1972), 597-605.

2. Ph. Brenner, The Cauchy problem for symmetric hyperbolic systems in L_p, Math. Scand. 19 (1966), 27-37.

3. Ph. Brenner, The Cauchy problem for systems in L_p and $L_{p,\alpha}$, Ark. Mat. 11 (1973), 75-101.

4. Ph. Brenner and V. Thomée, Stability and convergence rates in L_p for certain difference schemes, Math. Scand. 27 (1970), 5-23.

5. Ph. Brenner and V. Thomée, Estimates near discontinuities for some difference schemes. Math. Scand. 28 (1971), 329-340.

6. G.W. Hedstrom, Norms of powers of absolutely convergent Fourier series, Michigan Math. J. 13 (1966), 393-416.

7. G.W. Hedstrom, The rate of convergence of some difference schemes, SIAM J. Numer. Anal. 5 (1968), 363-406.

8. V. Thomée, On maximum-norm stable difference operators, Numerical Solution of Partial Differential Equations, Ed. J.H. Bramble, Academic Press, New York 1966, 125-151.

9. V. Thomée, On the rate of convergence of difference schemes for hyperbolic equations, Numerical Solution of Partial Differential Equations II, Ed. B. Hubbard, Academic Press, New York 1971, 585-622.

10. V. Thomée, Convergence analysis of a finite difference scheme for a simple semi-linear hyperbolic equation, Numerische Behandlung nichtlinearer Integro-differential- und Differentialgleichungen, Springer Lecture Notes in Mathematics 395, 149-166.

11. L. Wahlbin, Maximum norm estimates for Friedrichs' scheme in two dimensions, to appear in SIAM J. Numer. Anal. 11 (1974).

CHAPTER 6. THE SCHRÖDINGER EQUATION.

In this chapter we shall derive stability and convergence estimates in L_p for difference approximations to the Schrödinger equation. We first show in Section 1 that the solution of the initial value problem is bounded under certain smoothness assumptions in L_p on the initial data. In Section 2 we then determine the rate of growth in L_p of a difference operator E_k consistent with the initial value problem. In Section 3 we present precise L_p estimates in terms of the L_p-smoothness of the initial data for the rate of convergence of the finite difference approximation to the exact solution of the continuous problem. In Section 4 we prove some inverse results and finally, in Section 5, convergence estimates in the maximum norm, now measuring the smoothness of the initial function in L_1.

6.1. L_p-estimates for the initial value problem.

We shall consider the initial value problem

(1.1)
$$\frac{\partial u}{\partial t} = i \sum_{j=1}^{d} \frac{\partial^2 u}{\partial x_j^2} \equiv Pu, \quad \text{for} \quad x \in R^d,\ t > 0,$$
$$u(x,0) = v(x) .$$

In terms of the characteristic polynomial of P,

$$\hat{P}(\xi) = -i|\xi|^2 = -i \sum_{j=1}^{d} \xi_j^2 ,$$

the solution operator of (1.1) is defined by

$$E(t)v = \mathcal{F}^{-1}(\exp(t\hat{P})\hat{v}) = \mathcal{F}^{-1}(\exp(-it|\xi|^2)\hat{v}), \quad \text{for} \quad v \in \hat{C}_0^\infty .$$

We recall from Section 3.1 that since $|\exp(t\hat{P}(\xi))| = 1$ but $\exp(-i|\xi|^2) \notin M_p$ for $p \neq 2$, the initial value problem is well posed in L_2 but not in L_p for $p \neq 2$. Our purpose is now to prove boundedness in L_p of the solution $E(t)v$ under minimal smoothness assumptions on v in L_p.

<u>Theorem 1.1</u>. Let $1 \leq p \leq \infty$ and let $E(t)$ be the solution operator of (1.1). Then for each $T > 0$ there exists a constant C such that for $v \in B_p^{s',1}$,

$$\|E(t)v\|_p \leq C\|v\|_{B_p^{s',1}}, \text{ for } t \leq T,$$

where $s' = 2d|1/2 - 1/p|$.

<u>Proof</u>. By Lemma 2.6.1 it is sufficient to prove that with ψ_j the functions defined in Section 2.1,

$$(1.2) \qquad M_p(\psi_j \exp(t\hat{P})) \leq C2^{js'}, \text{ for } j \geq 0,\ t \leq T.$$

We shall show that for $g \in C_0^\infty$ fixed,

$$(1.3) \qquad M_p(g \exp(t\hat{P})) \leq C(1+t)^{s'/2} \text{ for } t > 0.$$

This clearly implies (1.2) for $j = 0$, and by a transformation of variables $\xi \to 2^j\xi$ also for $j > 0$, since then $\psi_j(\xi) = \phi(2^{-j}\xi)$.

Since g has compact support we obtain by Leibniz' formula

$$|D^\gamma(g(\xi)\exp(t\hat{P}(\xi)))| \leq C(1+t)^d, \text{ for } |\gamma| \leq d,$$

and hence by the Carlson-Beurling inequality,

$$M_\infty(g \exp(t\hat{P})) \leq C(1+t)^{d/2}.$$

Since also obviously

$$M_2(g \exp(t\hat{P})) = \|g\|_\infty = C,$$

(1.3) follows by interpolation (Theorem 1.2.5). This completes the proof of the theorem.

We now want to prove that the result of Theorem 1.1 is sharp with respect to the index s'. We shall first prove the following lemma:

Lemma 1.1. Let $1 \leq p \leq \infty$, and let $g \in C_0^\infty$, $g \not\equiv 0$. Then there exists a positive constant c such that with s' as in Theorem 1.1,

$$M_p(g \exp(t\hat{P})) \geq ct^{s'/2} \quad \text{for} \quad t > 0.$$

Proof. Let $\eta_k \not\equiv 0$, $k = 1,\ldots,d$, be functions in $C_0^\infty(R)$ such that $0 \notin \operatorname{supp}(\eta_k)$ and such that $\eta(\xi) = \prod_{k=1}^{d} \eta_k(\xi_k)$ has its support in the interior of $\operatorname{supp}(g)$. Then $\eta/g \in C_0^\infty(R^d) \subset M_p$ so that

$$(1.4) \qquad M_p(\eta \exp(t\hat{P})) \leq CM_p(g \exp(t\hat{P})).$$

On the other hand, by Lemma 1.2.2,

$$(1.5) \qquad M_p(\eta \exp(t\hat{P})) = \prod_{k=1}^{d} M_p^{(1)}(\eta_k \exp(-it\xi^2)),$$

and by Corollary 1.5.1 we have

$$(1.6) \qquad M_p^{(1)}(\eta_k \exp(-it\xi^2)) \geq ct^{\left|\frac{1}{2}-\frac{1}{p}\right|} \quad \text{for} \quad t > 0.$$

Together, (1.4), (1.5) and (1.6) prove the lemma.

We can now prove the sharpness of Theorem 1.1 with respect to s'.

Theorem 1.2. Let $1 \leq p \leq \infty$ and let $E(t)$ be the solution operator of (1.1). Assume that for some $t > 0$ there exists a constant C such that

$$(1.7) \qquad \|E(t)v\|_p \leq C\|v\|_{B_p^{s,1}}, \quad \text{for} \quad v \in \hat{C}_0^\infty.$$

Then $s \geq s'$.

Proof. By (1.7) we obtain for $v_j = \mathcal{F}^{-1}(\psi_j \hat{v})$, with constants throughout depending on t,

$$\|E(t)v_j\|_p \leq C\|v_j\|_{B_p^{s,1}} \leq C2^{js}\|v\|_p \quad \text{for} \quad j > 0.$$

Since for $j > 0$,

$$E(t)v_j = \mathcal{F}^{-1}(\phi_j \exp(t\hat{P})\hat{v}),$$

we conclude that

$$(1.8) \qquad M_p(\phi_j \exp(t\hat{P})) \leq C2^{js}.$$

On the other hand, we have by Lemma 1.1,

$$(1.9) \qquad M_p(\phi_j \exp(t\hat{P})) = M_p(\phi \exp(t2^{2j}\hat{P})) \geq c2^{js'}.$$

Together (1.8) and (1.9) imply that $s \geq s'$ which proves the theorem.

6.2. Growth estimates for finite difference operators.

We shall now derive estimates for the powers E_k^n of a finite difference operator consistent with (1.1),

$$(2.1) \qquad E_k v(x) = \mathcal{F}^{-1}(E_k v)(x) = \mathcal{F}^{-1}(e(h\cdot)\hat{v})(x) = \sum_{\beta \in Z^d} a_\beta v(x+h\beta), \quad k/h^2 = \lambda = \text{constant},$$

with characteristic function independent of h,

$$e(\xi) = \hat{E}_k(h^{-1}\xi) = \sum_{\beta \in Z^d} a_\beta e^{i\langle\beta,\xi\rangle}$$

(which it will again be sufficient to assume real analytic). Recall from Proposition 3.2.5 that since (1.1) is not well posed in L_p for $p \neq 2$, such a difference operator is unstable in L_p for $p \neq 2$, so that the operator norm $\|E_k^n\|_p$ is not uniformly bounded for $nk \leq T$.

We first prove the following general result valid for L_2 stable finite difference operators of the form (2.1) independent of their relation to any continuous problem.

Theorem 2.1. Let $1 \leq p \leq \infty$ and assume that E_k is stable in L_2. Then there exists a constant C such that

$$\|E_k^n\|_p \leq Cn^{d\left|\frac{1}{2}-\frac{1}{p}\right|}, \text{ for } n = 1,2,\ldots .$$

Proof. Since E_k is stable in L_2 we have $|e(\xi)| \leq 1$ and hence, since $e(\xi)$ is periodic and thus has bounded derivatives, we have

$$|D^\gamma e(\xi)^n| \leq Cn^{|\gamma|}, \text{ for } \xi \in R^d, |\gamma| \leq d.$$

With η as in Theorem 1.4.1 we therefore obtain by the Carlson-Beurling inequality,

$$M_\infty(e^n) \leq CM_\infty(\eta e^n) \leq Cn^{d/2},$$

which is the desired result for $p = \infty$. Since

$$M_2(e^n) = \|e^n\|_\infty \leq 1,$$

the result follows for general p by Theorem 1.2.5.

We shall now show that the estimate in Theorem 2.1 can be improved if instead of L_2 stability we make the stronger assumption that E_k is dissipative of order exactly σ, that is (for simplicity we assume that $e(\xi) \neq 0$ for $\xi \in R^d$),

(2.2) $$e(\xi) = \exp(-i\lambda|\xi|^2 + \rho(\xi)),$$

where $\rho(\xi) = o(|\xi|^2)$ as $\xi \to 0$, and with $c, c' > 0$,

$$-c'|\xi|^\sigma \leq \text{Re}\, \rho(\xi) \leq -c|\xi|^\sigma, \text{ for } |\xi_j| \leq \tfrac{3}{2}\pi, \; j = 1,\ldots,d .$$

Theorem 2.2. Let $1 \leq p \leq \infty$ and assume that E_k is consistent with (1.1) and dissipative of order exactly σ. Then

$$\|E_k^n\|_p \leq Cn^{d\left|\frac{1}{2}-\frac{1}{p}\right|(1-2/\sigma)} .$$

Proof. We shall prove that with η as in Theorem 1.4.1 we have

$$M_\infty(\eta e^n) \leq Cn^{\frac{1}{2}d(1-2/\sigma)} ,$$

from which the theorem follows for $p = \infty$ by Theorem 1.4.1 and, using also the L_2 stability, for general p by Theorem 1.2.5.

Let $\eta_n(\xi) = \eta(n^{-1/\sigma}\xi)$. Then

$$(2.3) \qquad M_\infty(\eta e^n) = M_\infty(\eta_n e(n^{-1/\sigma}\cdot)^n) .$$

In view of (2.2) we may write

$$e(n^{-1/\sigma}\xi)^n = \exp(-in^{1-2/\sigma}\theta_n(\xi))\exp(n \operatorname{Re} \rho(n^{-1/\sigma}\xi)) ,$$

where $\theta_n(\xi) = \lambda|\xi|^2 - n^{2/\sigma} \operatorname{Im}\rho(n^{-1/\sigma}\xi)$, so that for $\xi \in \operatorname{supp}(\eta_n)$ and $|\gamma| \leq d$,

$$(2.4) \qquad |D^\gamma\theta_n(\xi)| \leq C|\xi|^{\max(2-|\gamma|,0)} \leq C(1+|\xi|)^2,$$

and also, by the assumed dissipativity,

$$n \operatorname{Re} \rho(n^{-1/\sigma}\xi) \leq -c|\xi|^\sigma, \text{ with } c > 0,$$

$$(2.5)$$

$$|D^\gamma(n \operatorname{Re} \rho(n^{-1/\sigma}\xi))| \leq C|\xi|^{\max(\sigma-|\gamma|,0)} \leq C(1+|\xi|)^\sigma .$$

Let now $|\gamma| \leq d$. We may write $D^\gamma(\eta_n e(n^{-1/\sigma}\xi)^n)$ as a linear combination of terms of the form $D^{\gamma_1}\eta_n D^{\gamma_2}\exp(-in^{1-2/\sigma}\theta_n)D^{\gamma_3}\exp(n \operatorname{Re} \rho(n^{-1/\sigma}\xi))$, with $\gamma_1+\gamma_2+\gamma_3=\gamma$. The first factor here is uniformly bounded in n and ξ. Noticing that for $f = \exp(g)$, $D^\alpha f$ equals f times a sum of products of at most $|\alpha|$ derivatives of

g of orders at most $|\alpha|$, we find by (2.4) for the second factor,

$$|D^{\gamma_2}\exp(-in^{1-2/\sigma}\theta_n(\xi))| \leq Cn^{|\gamma_2|(1-2/\sigma)}(1+|\xi|)^{2|\gamma_2|} .$$

In the same way, using (2.5),

$$|D^{\gamma_3}\exp(n \operatorname{Re} \rho(n^{-1/\sigma}\xi))| \leq C(1+|\xi|)^{\sigma|\gamma_3|}\exp(-c|\xi|^{\sigma}) .$$

Since $|\gamma_j| \leq |\gamma|$, $j = 1,2,3$, we have

$$(2.6) \qquad |D^{\gamma}(\eta_n(\xi)e(n^{-1/\sigma}\xi)^n)| \leq Cn^{|\gamma|(1-2/\sigma)}(1+|\xi|)^{(2+\sigma)|\gamma|}\exp(-c|\xi|^{\sigma}).$$

The Carlson-Beurling inequality now proves

$$M_{\infty}(\eta_n e(n^{-1/\sigma}\cdot)^n) \leq Cn^{d(1-2/\sigma)/2} ,$$

which by (2.3) completes the proof of the theorem.

It follows from Theorem 5.3.1 that in the case of one space dimension ($d = 1$) the estimate of Theorem 2.1 is sharp if $|e(\xi)| \equiv 1$, and also that the estimate of Theorem 2.2 is sharp, since in that case the conditions (5.3.4) through (5.3.6) of Theorem 5.3.2 are satisfied with $\alpha = 0$, $\nu = 2$ and σ as above.

6.3. Convergence estimates in L_p.

In this section we shall prove convergence estimates in L_p for finite difference approximations to the initial value problem (1.1). The smoothness of the initial function will be measured in L_p. As in Section 2 we restrict ourselves to difference operators E_k with characteristic functions $e_k(\xi) = e(\xi)$ independent of k.

Recall that $s' = 2d|1/2 - 1/p|$. We first prove a finite difference analogue of Theorem 1.1.

Theorem 3.1. Let $1 \le p \le \infty$ and assume that E_k is consistent with (1.1) and stable in L_2. Then for each $T > 0$ there exists a constant C such that for $v \in B_p^{s',1}$,

$$\|E_k^n v\|_p \le C\|v\|_{B_p^{s',1}} \quad \text{for} \quad nk \le T.$$

Proof. We shall prove that

$$(3.1) \qquad M_\infty(\psi_j e(h\cdot)^n) \le C2^{jd} \quad \text{for} \quad nk \le T,\ j \ge 0.$$

Using the L_2 stability, it follows by Theorem 1.2.5 that

$$M_p(\psi_j e(h\cdot)^n) \le C2^{js'}$$

which in view of Lemma 2.6.1 proves the theorem.

In order to prove (3.1) we first notice that by consistency there is a positive δ such that we may write

$$e(\xi) = \exp(\theta(\xi)) \quad \text{for} \quad |\xi| < \delta ,$$

where

$$\theta(\xi) = -i\lambda|\xi|^2 + o(|\xi|^2) = O(|\xi|^2) \quad \text{as} \quad \xi \to 0.$$

Let $g \in C_0^\infty$ be fixed, and ω positive and so small that $\omega|\xi| < \delta$ for $\xi \in \text{supp}(g)$. Since then all derivatives of $\omega^{-2}\theta(\omega\xi)$ are bounded for $\xi \in \text{supp}(g)$, it follows by the argument used to prove (2.6) that for $|\gamma| \le d$,

$$|D^\gamma(g(\xi)e(\omega\xi)^n)| = |D^\gamma(g(\xi)\exp(n\omega^2(\omega^{-2}\theta(\omega\xi))))| \le C(1+n\omega^2)^{|\gamma|} ,$$

and hence by the Carlson-Beurling inequality, for these ω,

$$M_\infty(ge(\omega\cdot)^n) \le C(1+n\omega^2)^{d/2} .$$

In particular we obtain for small h,

$$(3.2) \qquad M_\infty(\psi_0 e(h\cdot)^n) \le C(1+nh^2)^{d/2} = C ,$$

and for $j > 0$ and $2^j h \leq \delta/2$,

$$(3.3) \qquad M_\infty(\psi_j e(h\cdot)^n) = M_\infty(\phi e(2^j h\xi)^n) \leq C[1+n(2^j h)^2]^{d/2} \leq C2^{jd}.$$

For j with $2^j h \geq \delta/2$, finally, we obtain by Theorem 2.1,

$$(3.4) \qquad M_\infty(\psi_j e(h\cdot)^n) \leq CM_\infty(e^n) \leq Cn^{d/2} = Ch^{-d} \leq C2^{jd}.$$

Together, (3.2), (3.3) and (3.4) prove (3.1) and hence complete the proof of the theorem.

We next prove an error estimate in L_p for smooth initial data. Recall that E_k is accurate of order μ if

$$e(\xi) = \exp(-i\lambda|\xi|^2 + O(|\xi|^{\mu+2})) \quad \text{as} \quad \xi \to 0.$$

Theorem 3.2. Let $1 \leq p \leq \infty$ and assume that E_k is consistent with (1.1), accurate of order μ, and stable in L_2. Then for each $T > 0$ there is a constant C such that for $v \in B_p^{2+\mu+s',1}$,

$$\|E_k^n v - E(nk)v\|_p \leq Ch^\mu \|v\|_{B_p^{2+\mu+s',1}} \quad \text{for} \quad nk \leq T.$$

Proof. Setting

$$(3.5) \qquad f_n(\xi) = e(\xi)^n - \exp(-in\lambda|\xi|^2),$$

the result will follow by Lemma 2.6.1 if we can prove that

$$(3.6) \qquad M_p(\psi_j f_n(h\cdot)) \leq Ch^\mu 2^{j(s'+\mu+2)} \quad \text{for} \quad nk \leq T,\ j \geq 0.$$

For $p = 2$ this is an immediate consequence of the fact that

$$|f_n(\xi)| \leq n|e(\xi) - \exp(-i\lambda|\xi|^2)| \leq Cn|\xi|^{\mu+2}, \text{ for } \xi \in R^d,$$

so that for $nk \leq T$,

$$M_2(\psi_j f_n(h\cdot)) = \|\psi_j f_n(h\cdot)\|_\infty \leq Cn(h2^j)^{\mu+2} \leq Ch^\mu 2^{j(\mu+2)} .$$

We shall prove that (3.6) holds also for $p = \infty$, from which the general result follows by Theorem 1.2.5.

Writing

$$(3.7) \qquad f_n(\xi) = \exp(-in\lambda|\xi|^2)r_n(\xi)$$

we have for small ξ,

$$(3.8) \qquad r_n(\xi) = \exp(n\rho(\xi)) - 1, \text{ with } \operatorname{Re}\rho \leq 0.$$

Here $\rho(\xi) = O(|\xi|^{\mu+2})$ as ξ tends to zero so that all derivatives of $\omega^{-(\mu+2)}\rho(\omega\xi)$ are bounded for $|\xi| \leq 2$ and $\omega > 0$ small, $\omega \leq \omega_0 \leq 1$, say. Hence we have for these ξ and ω and for $|\gamma| \leq d$,

$$|D^\gamma(r_n(\omega\xi))| = |D^\gamma(\exp(n\omega^{\mu+2}(\omega^{-(\mu+2)}\rho(\omega\xi))) - 1)| \leq C\{n\omega^{\mu+2} + (n\omega^{\mu+2})^d\} .$$

In particular, setting $\omega = 2^j h$, we obtain for j such that $2^{j(\mu+2)}h^\mu \leq \omega_0^\mu$ (≤ 1), that for $|\gamma| \leq d$ and $nk \leq T$,

$$(3.9) \qquad |D^\gamma r_n(2^j h\xi)| \leq Ch^\mu 2^{j(\mu+2)} .$$

Letting now $\tilde{\phi} = \phi$ for $j > 0$ and $\tilde{\phi} = \psi_0$ for $j = 0$, we obtain by the Carlson-Beurling inequality for $2^{j(\mu+2)}h^\mu \leq \omega_0^\mu$,

$$M_\infty(\tilde{\phi} r_n(2^j h\cdot)) \leq Ch^\mu 2^{j(\mu+2)} .$$

By (1.13) we have for any $g \in C_0^\infty$,

$$M_\infty(g \exp(-it2^{2j}|\xi|^2)) \leq C2^{jd} \quad \text{for } t \leq T,$$

and hence if in addition $g = 1$ for $|\xi| \le 2$, we conclude for $2^{j(\mu+2)}h^{\mu} \le \omega_0^{\mu}$, and $nk \le T$,

$$M_\infty(\psi_j f_n(h\cdot)) = M_\infty(\tilde{\phi} f_n(2^j h\cdot))$$

(3.10)

$$\le M_\infty(g \exp(-ink2^{2j}|\xi|^2)) M_\infty(\tilde{\phi} r_n(2^j h\cdot)) \le Ch^{\mu} 2^{j(d+\mu+2)} .$$

This proves the estimate (3.6) for $p = \infty$ and $2^{j(\mu+2)}h^{\mu} \le \omega_0^{\mu}$.

For j such that $2^{j(\mu+2)}h^{\mu} \ge \omega_0^{\mu}$, we find by (3.4) and (1.2) that

$$M_\infty(\psi_j f_n(h\cdot)) \le M_\infty(\psi_j e(h\cdot)^n) + M_\infty(\psi_j \exp(-ink|\xi|^2)) \le C2^{jd}$$

$$\le Ch^{\mu} 2^{j(d+\mu+2)} .$$

Together with (3.10) this completes the proof of (3.6) for $p = \infty$, and hence of the theorem.

By interpolation, Theorems 3.1 and 3.2 now yield the following result for less smooth initial data.

Theorem 3.3. Under the assumptions of Theorem 3.2, let $0 < s < \mu+2$. Then for each $T > 0$ there is a constant C such that for $v \in B_p^{s+s',\infty}$,

$$\|E_k^n v - E(nk)v\|_p \le Ch^{\frac{s\mu}{\mu+2}} \|v\|_{B_p^{s+s',\infty}} \quad \text{for } nk \le T.$$

6.4. Inverse results.

In this section we shall prove some inverse results similar to those of Sections 4.2 and 5.4. We begin by showing that in some sense the estimate of Theorem 3.3 is best possible.

Analogously to the parabolic and hyperbolic cases we say that the difference operator E_k is accurate of order exactly μ if for small ξ, $|\xi| \leq \delta \leq 1$ say, its characteristic function can be written

$$e(\xi) = \exp(-i\lambda|\xi|^2 + \rho(\xi)),$$

where with positive c and C,

$$c|\xi|^{\mu+2} \leq |\rho(\xi)| \leq C|\xi|^{\mu+2}. \tag{4.1}$$

Theorem 4.1. Let $1 \leq p \leq \infty$ and $s > 0$, and assume that E_k is consistent with (1.1), accurate of order exactly μ, and stable in L_2. Then for each $t > 0$ there exist positive constants c and h_0 such that for $h \leq h_0$, $nk = t$,

$$\sup\{\|E_k^n v - E(nk)v\|_p : v \in B_p^{s+s',\infty}, \|v\|_{B_p^{s+s',\infty}} \leq 1\} \geq ch^{q(s)}, \tag{4.2}$$

where $s' = 2d|1/2 - 1/p|$ and $q(s) = \min(\mu, s\mu/(\mu+2))$.

Proof. With the notation (3.5) and $\bar{s} = s + s'$, (4.2) may be written

$$\sup\{\|\mathcal{F}^{-1}(f_n(h\cdot)\hat{v})\|_p : v \in \hat{C}_0^\infty, \|v\|_{B_p^{\bar{s},\infty}} \leq 1\} \geq ch^{q(s)}.$$

Taking the supremum only over functions of the form $v = \mathcal{F}^{-1}(\phi_j\hat{w})$, and noticing that for $j \geq 0$,

$$\|\mathcal{F}^{-1}(\phi_j\hat{w})\|_{B_p^{\bar{s},\infty}} \leq C2^{\bar{s}j}\|w\|_p,$$

we find that it suffices to show that for $h \leq h_0$, $nk = t$,

$$\sup_{j \geq 0} 2^{-\bar{s}j} M_p(f_n(h\cdot)\phi_j)$$

$$= \sup\{\|\mathcal{F}^{-1}(f_n(h\cdot)\phi_j\hat{w})\|_p : w \in \hat{C}_0^\infty, \|w\|_p \leq 2^{-\bar{s}j}, j \geq 0\} \geq ch^{q(s)},$$

or

$$(4.3) \qquad \sup_{j \geq 0} 2^{-\bar{s}j} M_p(f_n(2^j h\cdot)\phi) \geq ch^{q(s)}, \text{ for } h \leq h_0,\ nk = t\,.$$

For $|\xi| \leq \delta$ we shall use r_n defined by (3.7), (3.8). Setting $\omega = 2^j h$, we find, if ε is sufficiently small (with $2\varepsilon < \delta$) that for $\frac{1}{2} < |\xi| < 2$, $\omega \leq \varepsilon$, $n\omega^{\mu+2} \leq \varepsilon$, we have in view of (4.1),

$$|r_n(\omega\xi)| \geq cn\omega^{\mu+2}.$$

Further, we obtain similarly to (3.9) for these ξ, ω and n,

$$|D^\gamma(\phi(\xi)r_n(\omega\xi))| \leq Cn\omega^{\mu+2}, \text{ for } |\gamma| \leq d,$$

and since $D^\gamma(r_n(\omega\xi)^{-1})$ equals $r_n(\omega\xi)^{-1}$ times a linear combination of products of factors of the form $r_n(\omega\xi)^{-1}D^{\gamma'}(r_n(\omega\xi))$, with $|\gamma'| \leq |\gamma|$, it follows for $\omega \leq \varepsilon$, $n\omega^{\mu+2} \leq \varepsilon$,

$$|D^\gamma(\phi(\xi)r_n(\omega\xi)^{-1})| \leq C(n\omega^{\mu+2})^{-1}\,.$$

Hence the Carlson-Beurling inequality shows that

$$(4.4) \qquad M_p(\phi r_n(\omega\cdot)^{-1}) \leq M_\infty(\phi r_n(\omega\cdot)^{-1}) \leq C(n\omega^{\mu+2})^{-1}\,.$$

Using also Lemma 1.1, we may therefore conclude, for these n and $\omega = h2^j$,

$$c2^{js'} \leq M_p(\phi^2\exp(-it2^{2j}|\xi|^2)) = M_p(\phi f_n(\omega\cdot)\phi r_n(\omega\cdot)^{-1})$$
$$\leq CM_p(\phi f_n(\omega\cdot))(n\omega^{\mu+2})^{-1}\,,$$

so that

$$(4.5) \qquad 2^{-js'}M_p(\phi f_n(\omega\cdot)) \geq ch^\mu 2^{j(\mu+2-s)}\,.$$

If $s \geq \mu+2$, we take $j = 0$ and h_0 so small that $h_0 \leq \varepsilon$, $nh_0^{\mu+2} \leq \varepsilon$, and obtain

$$M_p(\phi f_n(\omega\cdot)) \geq ch^{\mu} \quad \text{for} \quad h \leq h_0, \ nk = t,$$

which proves (4.3) in this case. If $0 < s < \mu+2$, we let j be the largest integer such that $2^{j(\mu+2)}h^{\mu} \leq \varepsilon$. For sufficiently small h, j is non-negative and it follows that $\omega = h2^j \leq (\varepsilon 2^{-2j})^{1/\mu} \leq \varepsilon$ and hence from (4.5),

$$2^{-j\bar{s}}M_p(\phi f_n(\omega\cdot)) \geq ch^{\mu}(h^{-\mu/(\mu+2)})^{\mu+2-s} = ch^{s\mu/(\mu+2)} .$$

This proves (4.3) for these s and thus completes the proof of the theorem.

We shall now prove that from a known rate of convergence (for simplicity here in the maximum norm) certain conclusions may be drawn about the smoothness of the initial data.

<u>Theorem 4.2</u>. Under the assumptions of Theorem 4.1, assume in addition that $v \in L_\infty$, that $s > d$, and that for $nk = t > 0$ fixed,

$$\|E_k^n v - E(nk)v\|_\infty = O(h^{\frac{s\mu}{\mu+2}}) \quad \text{as} \quad h \to 0.$$

Then $v \in B_\infty^{s-d,\infty}$.

<u>Proof</u>. It is enough to prove that for some j_0,

$$\|\mathcal{F}^{-1}(\phi_j\hat{v})\|_\infty \leq C2^{-j(s-d)} \quad \text{for} \quad j \geq j_0.$$

Without loss of generality we may assume that $t = 1$.

Let $\omega = 2^j h \leq \varepsilon$ and $n\omega^{\mu+2} = 2^{j(\mu+2)}h^{\mu}/\lambda \leq \varepsilon$. We then have using the notation (3.5), (3.7) and (3.8)

$$\mathcal{F}^{-1}(\phi_j\hat{v}) = \mathcal{F}^{-1}(\exp(i|\xi|^2)r_n(h\xi)^{-1}f_n(h\xi)\phi_j\hat{v})$$

$$= \mathcal{F}^{-1}(\exp(i|\xi|^2)r_n(h\xi)^{-1}\phi_j\mathcal{F}(E_k^n v - E(1)v)) .$$

Let $g \in C_0^\infty$ be such that $g = 1$ on $\mathrm{supp}(\phi)$ and set $g_j(\xi) = g(2^{-j}\xi)$. Then $g_j\phi_j = \phi_j$, and hence if j is such that $n\omega^{\mu+2} \leq \varepsilon$, $\omega \leq \varepsilon$,

$$(4.6) \qquad \|\mathcal{F}^{-1}(\phi_j\hat{v})\|_\infty \leq CM_\infty(g_j \exp(i|\xi|^2))M_\infty(\phi_j r_n(h\cdot)^{-1})h^{s\mu/(\mu+2)} .$$

Choosing $h = h(j)$ so that $n = 1/(\lambda h^2) \in Z$ and $\varepsilon/2 \leq 2^{j(\mu+2)}h^\mu/\lambda \leq \varepsilon$, which is possible if j is large, we have by (4.4) (noticing that then for large j, $\omega \leq (\varepsilon\lambda 2^{-2j})^{1/\mu} \leq \varepsilon$),

$$M_\infty(\phi_j r_n(h\cdot)^{-1}) = M_\infty(\phi r_n(\omega\cdot)^{-1}) \leq C,$$

while the Carlson-Beurling inequality shows that

$$M_\infty(g_j \exp(i|\xi|^2)) = M_\infty(g \exp(i2^{2j}|\xi|^2)) \leq C2^{jd} .$$

By (4.6) this implies that for large j,

$$\|\mathcal{F}^{-1}(\phi_j\hat{v})\|_\infty \leq C2^{jd}h^{\frac{s\mu}{\mu+2}} \leq C2^{-j(s-d)} ,$$

and the theorem is proved.

Notice the gap between the positive convergence result of Theorem 3.3, and the inverse result of Theorem 4.2. In order to obtain an $O(h^{s\mu/(\mu+2)})$ convergence estimate in the maximum norm we need to assume that the initial function is in $B_\infty^{s+d,\infty}$, whereas conversely from this convergence rate only follows that the initial function is in $B_\infty^{s-d,\infty}$. We shall return to comment on this at the end of the next section.

6.5. Convergence estimates from L_1 to L_∞.

In somewhat the same way as for the heat equation in Section 4.3, it is possible, for t bounded away from zero, to derive maximum norm error estimates with the smoothness of the data measured in L_1, and thus improve in certain cases the estimates obtained in Section 3.

For simplicity we consider now only difference schemes for the initial value problem for the one-dimensional Schrödinger equation,

$$\frac{\partial u}{\partial t} = i\,\frac{\partial^2 u}{\partial x^2}\,,\ \text{for}\ \ x \in R^1,\ t > 0, \tag{5.1}$$

$$u(x,0) = v(x)\,.$$

We begin with a result with maximal order of convergence.

<u>Theorem 5.1</u>. Assume that E_k is consistent with (5.1), accurate of order μ and stable in L_2. Then for t_0,T with $0 < t_0 < T$ there exists a constant C such that for $v \in B_1^{\mu+2,1}$,

$$\|E_k^n v - E(nk)v\|_\infty \le Ch^\mu \|v\|_{B_1^{\mu+2,1}},\ \text{for}\ \ t_0 \le nk \le T.$$

<u>Proof</u>. Introducing again $v_j = \mathcal{F}^{-1}(\phi_j\hat{v})$, we may write for $v \in \hat{C}_0^\infty$, with f_n as in (3.5),

$$F_{nk}v = E_k^n v - E(nk)v = \mathcal{F}^{-1}(f_n(h\cdot)\hat{v}) = \sum_{j=-\infty}^{\infty}\ \sum_{l=j-1}^{j+1} \mathcal{F}^{-1}(f_n(h\cdot)\phi_j\hat{v}_l)$$

$$= \sum_{j=-\infty}^{\infty}\ \sum_{l=j-1}^{j+1} \mathcal{F}^{-1}(f_n(h\cdot)\phi_j) * v_l\,,$$

and hence

$$\|F_{nk}v\|_\infty \le \sum_{j=-\infty}^{\infty}\ \sum_{l=j-1}^{j+1} \|\mathcal{F}^{-1}(f_n(h\cdot)\phi_j)\|_\infty\, \|v_l\|_1\,. \tag{5.2}$$

The theorem will now follow from

$$\|\mathcal{F}^{-1}(f_n(h\xi)\phi_j)\|_\infty \le Ch^\mu 2^{j(\mu+2)} \quad \text{for}\ \ j \in Z. \tag{5.3}$$

In the notation (3.7), (3.8) we have for $|\xi| \leq \delta$, say,

$$|(\frac{d}{d\xi})^l r_n(\xi)| \leq n|(\frac{d}{d\xi})^l \rho(\xi)| \leq Cn|\xi|^{\mu+2-l}, \quad l = 0,1,$$

while for $|\xi| \geq \delta$ it follows for $r_n(\xi) = \exp(in\lambda\xi^2)e(\xi)^n - 1$, since $e(\xi)$ and $\frac{d}{d\xi} e(\xi)$ are bounded, that

$$|(\frac{d}{d\xi})^l r_n(\xi)| \leq C(n|\xi|)^l \leq Cn|\xi|^{\mu+2-l}, \quad l = 0,1.$$

By Lemma 1.5.2 we obtain then for $0 < t_0 < t = nk \leq T$,

$$\|\mathcal{F}^{-1}(f_n(h\xi)\phi_j)\|_\infty = 2^j\|\mathcal{F}^{-1}(\exp(-it2^{2j}\xi^2)r_n(2^jh\xi)\phi)\|_\infty$$

$$\leq C\|\frac{d}{d\xi}(r_n(2^jh\xi)\phi)\|_1 \leq Ch^\mu 2^{j(\mu+2)},$$

which proves (5.3) and hence completes the proof of the theorem.

We next turn to the estimate for less smooth v.

<u>Theorem 5.2</u>. Assume that E_k is consistent with (5.1), accurate of order μ and stable in L_2. Let $1 < s < \mu+2$, $s \neq (\mu+2)/2$, and $0 < t_0 < T$. Then there exists a constant C such that for $v \in B_1^{s,\infty}$,

$$\|E_k^n v - E(nk)v\|_\infty \leq Ch^{q(s)}\|v\|_{B_1^{s,\infty}} \quad \text{for } t_0 \leq nk \leq T,$$

where

$$q(s) = \begin{cases} s\mu/(\mu+2), & \text{for } \frac{1}{2}(\mu+2) < s < \mu+2, \\ s-1, & \text{for } 1 < s < \frac{1}{2}(\mu+2). \end{cases}$$

<u>Proof</u>. We shall prove that with $F_{nk} = E_k^n - E(nk)$,

(5.4) $$\|F_{nk}v\|_\infty \leq Ch^{\mu/2}\|v\|_{\dot{B}_1^{(\mu+2)/2,1}},$$

and

$$(5.5) \qquad \|F_{nk}v\|_\infty \le C\|v\|_{B_1^{1,1}} .$$

The result then follows by interpolation (Corollary 2.5.1) between (5.4) and (5.5) for $1 < s < (\mu+2)/2$ and between (5.4) and Theorem 5.1 for $(\mu+2)/2 < s < \mu+2$.

In order to prove (5.4) it suffices by (5.2) to prove that

$$(5.6) \qquad \|\mathcal{F}^{-1}(f_n(h\xi)\phi_j)\|_\infty \le Ch^{\mu/2}2^{(\mu+2)j/2} \quad \text{for } j \in Z,$$

and it is easy to see that (5.5) similarly follows from

$$(5.7) \qquad \|\mathcal{F}^{-1}(f_n(h\xi)\psi_j)\|_\infty \le C2^j, \quad j = 0,1,\dots .$$

In order to prove (5.7) we need only notice that since $|f_n(\xi)| \le 2$,

$$(5.8) \qquad \|\mathcal{F}^{-1}(f_n(h\xi)\psi_j)\|_\infty \le 2\|\psi_j\|_1 \le C2^j \quad \text{for } j \ge 0.$$

It remains to prove (5.6). For $h^\mu 2^{j(\mu+2)} \le 1$ this is an immediate consequence of (5.3) and it therefore suffices to consider $h^\mu 2^{j(\mu+2)} \ge 1$. By a change of variables and van der Corput's lemma (Lemma 1.5.2) we then obtain for $t \ge t_0$,

$$(5.9) \qquad \|\mathcal{F}^{-1}(\exp(-it\xi^2)\phi_j)\|_\infty = 2^j\|\mathcal{F}^{-1}(\exp(-it2^{2j}\xi^2)\phi)\|_\infty \le C \le Ch^{\mu/2}2^{(\mu+2)j/2} ,$$

and it remains to prove a similar estimate for $\mathcal{F}^{-1}(e(h\xi)^n\phi_j)$. We shall prove below that then there exists an $\varepsilon > 0$ such that

$$(5.10) \qquad \|\mathcal{F}^{-1}(e(h\xi)^n\phi_j)\|_\infty \le C \quad \text{for } 2^j h \le \varepsilon,\ t_0 \le nk \le T.$$

From this we obtain at once for $h2^j \le \varepsilon$ but $h^\mu 2^{j(\mu+2)} \ge 1$,

$$(5.11) \qquad \|\mathcal{F}^{-1}(e(h\xi)^n\phi_j)\|_\infty \le Ch^{\mu/2}2^{(\mu+2)j/2} .$$

For the remaining possibility, $h2^j > \varepsilon$, we have as in (5.8),

(5.12) $$\|\mathcal{F}^{-1}(e(h\xi)^n\phi_j)\|_\infty \le \|\phi_j\|_1 \le C2^j \le C2^j(h2^j/\varepsilon)^{\mu/2} = Ch^{\mu/2}2^{(\mu+2)j/2} .$$

Together, (5.9), (5.11) and (5.12) prove (5.6) for $h^\mu 2^{j(\mu+2)} \ge 1$.

In order, finally, to prove (5.10), we write for $|\xi| \le \delta$,

$$e(\xi) = \exp(-i\lambda\xi^2 + i \operatorname{Im} \rho(\xi))\exp(\operatorname{Re} \rho(\xi)).$$

Let $\theta(\xi) = -\lambda\xi^2 + \operatorname{Im} \rho(\xi)$. Then choosing ε sufficiently small with $2\varepsilon \le \delta$, we have for $0 < \omega \le \varepsilon$ and $1/2 \le |\xi| \le 2$,

$$-\frac{d^2}{d\xi^2}(\omega^{-2}\theta(\omega\xi)) \ge c > 0,$$

and since $\rho(\xi)$ is analytic for $|\xi| \le 2\varepsilon$, and by the L_2 stability, either $\operatorname{Re}\rho(\xi) \equiv 0$ or else there exist positive constants $\gamma, \gamma', \gamma''$ and an even integer $\sigma \ge \mu+2$ such that for $|\xi| \le 2\varepsilon$,

$$-\gamma'\xi^\sigma \le \operatorname{Re} \rho(\xi) \le -\gamma\xi^\sigma, \ |\operatorname{Re} \rho'(\xi)| \le \gamma''|\xi|^{\sigma-1} .$$

In either case we then have for $\omega = h2^j \le \varepsilon$,

$$\left|\frac{d}{d\xi}(\exp(n \operatorname{Re} \rho(\omega\xi))\phi(\xi))\right| \le C, \quad nk \le T.$$

Hence, using Lemma 1.5.2 we have for $\omega = h2^j \le \varepsilon$,

$$\|\mathcal{F}^{-1}(e(h\cdot)^n\phi_j)\|_\infty = 2^j\|\mathcal{F}^{-1}(e(\omega\cdot)^n\phi)\|_\infty$$

$$= 2^j\|\mathcal{F}^{-1}(\exp(in\omega^2(\omega^{-2}\theta(\omega\cdot)))\exp(n \operatorname{Re} \rho(\omega\cdot))\phi)\|_\infty$$

$$\le C2^j(n\omega^2)^{-1/2} \left\|\frac{d}{d\xi}(\exp(n \operatorname{Re} \rho(\omega\cdot))\phi)\right\|_1 \le C(nk)^{-1/2} = C,$$

which proves (5.10) and the proof of the theorem is completed.

As an example, we apply our different convergence estimates in the maximum norm to the piecewise smooth function H_s of Example II of Section 2.4. Recall that H_s belongs to $B_p^{r,\infty}$ if and only if $r \le s + 1/p$ for $1 \le p \le \infty$. From Theorem 3.3.8 we then obtain an error estimate of $O(h^{s\mu/(\mu+2)})$ as h tends to zero for $0 < s < \mu+2$, and Theorem 1.3 yields $O(h^{(s-1)\mu/(\mu+2)})$, for $1 < s < \mu+3$. Finally, for $t = nk$ bounded away from zero our last result, Theorem 5.2, gives $O(h^{(s+1)\mu/(\mu+2)})$ for $\mu/2 < s < \mu+1$, which is best possible in view of Theorem 4.2, and $O(h^s)$ for $0 < s < \mu/2$.

References.

For the initial value problem, results similar to those of Section 1 were obtained in [2]. For difference schemes, a number of the results presented in Sections 2 through 4 can be found in [1].

1. G. Fernström, Convergence in L_p of discrete solutions to the Schrödinger equation, Report 1971-13, Department of Mathematics, Chalmers Institute of Technology and the University of Göteborg, Göteborg, Sweden.

2. E. Lanconelli, Valutazione in $L_p(R^n)$ della solutione del problema di Cauchy per l'equazione di Schrödinger, Boll. Un. Mat. Ital. 4(1968), 591-607.

INDEX

FREQUENTLY USED SPACES

Vol. 277: Séminaire Banach. Edité par C. Houzel. VII, 229 pages. 1972. DM 20,–

Vol. 278: H. Jacquet, Automorphic Forms on GL(2). Part II. XIII, 142 pages. 1972. DM 16,–

Vol. 279: R. Bott, S. Gitler and I. M. James, Lectures on Algebraic and Differential Topology. V, 174 pages. 1972. DM 18,–

Vol. 280: Conference on the Theory of Ordinary and Partial Differential Equations. Edited by W. N. Everitt and B. D. Sleeman. XV, 367 pages. 1972. DM 26,–

Vol. 281: Coherence in Categories. Edited by S. Mac Lane. VII, 235 pages. 1972. DM 20,–

Vol. 282: W. Klingenberg und P. Flaschel, Riemannsche Hilbertmannigfaltigkeiten. Periodische Geodätische. VII, 211 Seiten. 1972. DM 20,–

Vol. 283: L. Illusie, Complexe Cotangent et Déformations II. VII, 304 pages. 1972. DM 24,–

Vol. 284: P. A. Meyer, Martingales and Stochastic Integrals I. VI, 89 pages. 1972. DM 16,–

Vol. 285: P. de la Harpe, Classical Banach-Lie Algebras and Banach-Lie Groups of Operators in Hilbert Space. III, 160 pages. 1972. DM 16,–

Vol. 286: S. Murakami, On Automorphisms of Siegel Domains. V, 95 pages. 1972. DM 16,–

Vol. 287: Hyperfunctions and Pseudo-Differential Equations. Edited by H. Komatsu. VII, 529 pages. 1973. DM 36,–

Vol. 288: Groupes de Monodromie en Géométrie Algébrique. (SGA 7 I). Dirigé par A. Grothendieck. IX, 523 pages. 1972. DM 50,–

Vol. 289: B. Fuglede, Finely Harmonic Functions. III, 188. 1972. DM 18,–

Vol. 290: D. B. Zagier, Equivariant Pontrjagin Classes and Applications to Orbit Spaces. IX, 130 pages. 1972. DM 16,–

Vol. 291: P. Orlik, Seifert Manifolds. VIII, 155 pages. 1972. DM 16,–

Vol. 292: W. D. Wallis, A. P. Street and J. S. Wallis, Combinatorics: Room Squares, Sum-Free Sets, Hadamard Matrices. V, 508 pages. 1972. DM 50,–

Vol. 293: R. A. DeVore, The Approximation of Continuous Functions by Positive Linear Operators. VIII, 289 pages. 1972. DM 24,–

Vol. 294: Stability of Stochastic Dynamical Systems. Edited by R. F. Curtain. IX, 332 pages. 1972. DM 26,–

Vol. 295: C. Dellacherie, Ensembles Analytiques, Capacités, Mesures de Hausdorff. XII, 123 pages. 1972. DM 16,–

Vol. 296: Probability and Information Theory II. Edited by M. Behara, K. Krickeberg and J. Wolfowitz. V, 223 pages. 1973. DM 20,–

Vol. 297: J. Garnett, Analytic Capacity and Measure. IV, 138 pages. 1972. DM 16,–

Vol. 298: Proceedings of the Second Conference on Compact Transformation Groups. Part 1. XIII, 453 pages. 1972. DM 32,–

Vol. 299: Proceedings of the Second Conference on Compact Transformation Groups. Part 2. XIV, 327 pages. 1972. DM 26,–

Vol. 300: P. Eymard, Moyennes Invariantes et Représentations Unitaires. II. 113 pages. 1972. DM 16,–

Vol. 301: F. Pittnauer, Vorlesungen über asymptotische Reihen. VI, 186 Seiten. 1972. DM 18,–

Vol. 302: M. Demazure, Lectures on p-Divisible Groups. V, 98 pages. 1972. DM 16,–

Vol. 303: Graph Theory and Applications. Edited by Y. Alavi, D. R. Lick and A. T. White. IX, 329 pages. 1972. DM 26,–

Vol. 304: A. K. Bousfield and D. M. Kan, Homotopy Limits, Completions and Localizations. V, 348 pages. 1972. DM 26,–

Vol. 305: Théorie des Topos et Cohomologie Etale des Schémas. Tome 3. (SGA 4). Dirigé par M. Artin, A. Grothendieck et J. L. Verdier. VI, 640 pages. 1973. DM 50,–

Vol. 306: H. Luckhardt, Extensional Gödel Functional Interpretation. VI, 161 pages. 1973. DM 18,–

Vol. 307: J. L. Bretagnolle, S. D. Chatterji et P.-A. Meyer, Ecole d'été de Probabilités: Processus Stochastiques. VI, 198 pages. 1973. DM 20,–

Vol. 308: D. Knutson, λ-Rings and the Representation Theory of the Symmetric Group. IV, 203 pages. 1973. DM 20,–

Vol. 309: D. H. Sattinger, Topics in Stability and Bifurcation Theory. VI, 190 pages. 1973. DM 18,–

Vol. 310: B. Iversen, Generic Local Structure of the Morphisms in Commutative Algebra. IV, 108 pages. 1973. DM 16,–

Vol. 311: Conference on Commutative Algebra. Edited by J. W. Brewer and E. A. Rutter. VII, 251 pages. 1973. DM 22,–

Vol. 312: Symposium on Ordinary Differential Equations. Edited by W. A. Harris, Jr. and Y. Sibuya. VIII, 204 pages. 1973. DM 22,–

Vol. 313: K. Jörgens and J. Weidmann, Spectral Properties of Hamiltonian Operators. III, 140 pages. 1973. DM 16,–

Vol. 314: M. Deuring, Lectures on the Theory of Algebraic Functions of One Variable. VI, 151 pages. 1973. DM 16,–

Vol. 315: K. Bichteler, Integration Theory (with Special Attention to Vector Measures). VI, 357 pages. 1973. DM 26,–

Vol. 316: Symposium on Non-Well-Posed Problems and Logarithmic Convexity. Edited by R. J. Knops. V, 176 pages. 1973. DM 18,–

Vol. 317: Séminaire Bourbaki – vol. 1971/72. Exposés 400–417. IV, 361 pages. 1973. DM 26,–

Vol. 318: Recent Advances in Topological Dynamics. Edited by A. Beck, VIII, 285 pages. 1973. DM 24,–

Vol. 319: Conference on Group Theory. Edited by R. W. Gatterdam and K. W. Weston. V, 188 pages. 1973. DM 18,–

Vol. 320: Modular Functions of One Variable I. Edited by W. Kuyk. V, 195 pages. 1973. DM 18,–

Vol. 321: Séminaire de Probabilités VII. Edité par P. A. Meyer. VI, 322 pages. 1973. DM 26,–

Vol. 322: Nonlinear Problems in the Physical Sciences and Biology. Edited by I. Stakgold, D. D. Joseph and D. H. Sattinger. VIII, 357 pages. 1973. DM 26,–

Vol. 323: J. L. Lions, Perturbations Singulières dans les Problèmes aux Limites et en Contrôle Optimal. XII, 645 pages. 1973. DM 42,–

Vol. 324: K. Kreith, Oscillation Theory. VI, 109 pages. 1973. DM 16,–

Vol. 325: Ch.-Ch. Chou, La Transformation de Fourier Complexe et L'Equation de Convolution. IX, 137 pages. 1973. DM 16,–

Vol. 326: A. Robert, Elliptic Curves. VIII, 264 pages. 1973. DM 22,–

Vol. 327: E. Matlis, 1-Dimensional Cohen-Macaulay Rings. XII, 157 pages. 1973. DM 18,–

Vol. 328: J. R. Büchi and D. Siefkes, The Monadic Second Order Theory of All Countable Ordinals. VI, 217 pages. 1973. DM 20,–

Vol. 329: W. Trebels, Multipliers for (C, α)-Bounded Fourier Expansions in Banach Spaces and Approximation Theory. VII, 103 pages. 1973. DM 16,–

Vol. 330: Proceedings of the Second Japan-USSR Symposium on Probability Theory. Edited by G. Maruyama and Yu. V. Prokhorov. VI, 550 pages. 1973. DM 36,–

Vol. 331: Summer School on Topological Vector Spaces. Edited by L. Waelbroeck. VI, 226 pages. 1973. DM 20,–

Vol. 332: Séminaire Pierre Lelong (Analyse) Année 1971-1972. V, 131 pages. 1973. DM 16,–

Vol. 333: Numerische, insbesondere approximationstheoretische Behandlung von Funktionalgleichungen. Herausgegeben von R. Ansorge und W. Törnig. VI, 296 Seiten. 1973. DM 24,–

Vol. 334: F. Schweiger, The Metrical Theory of Jacobi-Perron Algorithm. V, 111 pages. 1973. DM 16,–

Vol. 335: H. Huck, R. Roitzsch, U. Simon, W. Vortisch, R. Walden, B. Wegner und W. Wendland, Beweismethoden der Differentialgeometrie im Großen. IX, 159 Seiten. 1973. DM 18,–

Vol. 336: L'Analyse Harmonique dans le Domaine Complexe. Edité par E. J. Akutowicz. VIII, 169 pages. 1973. DM 18,–

Vol. 337: Cambridge Summer School in Mathematical Logic. Edited by A. R. D. Mathias and H. Rogers. IX, 660 pages. 1973. DM 42,–

Vol: 338: J. Lindenstrauss and L. Tzafriri, Classical Banach Spaces. IX, 243 pages. 1973. DM 22,–

Vol. 339: G. Kempf, F. Knudsen, D. Mumford and B. Saint-Donat, Toroidal Embeddings I. VIII, 209 pages. 1973. DM 20,–

Vol. 340: Groupes de Monodromie en Géométrie Algébrique. (SGA 7 II). Par P. Deligne et N. Katz. X, 438 pages. 1973. DM 40,–

Vol. 341: Algebraic K-Theory I, Higher K-Theories. Edited by H. Bass. XV, 335 pages. 1973. DM 26,–

Vol. 342: Algebraic K-Theory II, "Classical" Algebraic K-Theory, and Connections with Arithmetic. Edited by H. Bass. XV, 527 pages. 1973. DM 36,–

Vol. 343: Algebraic K-Theory III, Hermitian K-Theory and Geometric Applications. Edited by H. Bass. XV, 572 pages. 1973. DM 38,–

Vol. 344: A. S. Troelstra (Editor), Metamathematical Investigation of Intuitionistic Arithmetic and Analysis. XVII, 485 pages. 1973. DM 34,–

Vol. 345: Proceedings of a Conference on Operator Theory. Edited by P. A. Fillmore. VI, 228 pages. 1973. DM 20,–

Vol. 346: Fučík et al., Spectral Analysis of Nonlinear Operators. II, 287 pages. 1973. DM 26,–

Vol. 347: J. M. Boardman and R. M. Vogt, Homotopy Invariant Algebraic Structures on Topological Spaces. X, 257 pages. 1973. DM 22,–

Vol. 348: A. M. Mathai and R. K. Saxena, Generalized Hypergeometric Functions with Applications in Statistics and Physical Sciences. VII, 314 pages. 1973. DM 26,–

Vol. 349: Modular Functions of One Variable II. Edited by W. Kuyk and P. Deligne. V, 598 pages. 1973. DM 38,–

Vol. 350: Modular Functions of One Variable III. Edited by W. Kuyk and J.-P. Serre. V, 350 pages. 1973. DM 26,–

Vol. 351: H. Tachikawa, Quasi-Frobenius Rings and Generalizations. XI, 172 pages. 1973. DM 18,–

Vol. 352: J. D. Fay, Theta Functions on Riemann Surfaces. V, 137 pages. 1973. DM 16,–

Vol. 353: Proceedings of the Conference. on Orders, Group Rings and Related Topics. Organized by J. S. Hsia, M. L. Madan and T. G. Ralley. X, 224 pages. 1973. DM 20,–

Vol. 354: K. J. Devlin, Aspects of Constructibility. XII, 240 pages. 1973. DM 22,–

Vol. 355: M. Sion, A Theory of Semigroup Valued Measures. V, 140 pages. 1973. DM 16,–

Vol. 356: W. L. J. van der Kallen, Infinitesimally Central-Extensions of Chevalley Groups. VII, 147 pages. 1973. DM 16,–

Vol. 357: W. Borho, P. Gabriel und R. Rentschler, Primideale in Einhüllenden auflösbarer Lie-Algebren. V, 182 Seiten. 1973. DM 18,–

Vol. 358: F. L. Williams, Tensor Products of Principal Series Representations. VI, 132 pages. 1973. DM 16,–

Vol. 359: U. Stammbach, Homology in Group Theory. VIII, 183 pages. 1973. DM 18,–

Vol. 360: W. J. Padgett and R. L. Taylor, Laws of Large Numbers for Normed Linear Spaces and Certain Fréchet Spaces. VI, 111 pages. 1973. DM 16,–

Vol. 361: J. W. Schutz, Foundations of Special Relativity: Kinematic Axioms for Minkowski Space Time. XX, 314 pages. 1973. DM 26,–

Vol. 362: Proceedings of the Conference on Numerical Solution of Ordinary Differential Equations. Edited by D. Bettis. VIII, 490 pages. 1974. DM 34,–

Vol. 363: Conference on the Numerical Solution of Differential Equations. Edited by G. A. Watson. IX, 221 pages. 1974. DM 20,–

Vol. 364: Proceedings on Infinite Dimensional Holomorphy. Edited by T. L. Hayden and T. J. Suffridge. VII, 212 pages. 1974. DM 20,–

Vol. 365: R. P. Gilbert, Constructive Methods for Elliptic Equations. VII, 397 pages. 1974. DM 26,–

Vol. 366: R. Steinberg, Conjugacy Classes in Algebraic Groups (Notes by V. V. Deodhar). VI, 159 pages. 1974. DM 18,–

Vol. 367: K. Langmann und W. Lütkebohmert, Cousinverteilungen und Fortsetzungssätze. VI, 151 Seiten. 1974. DM 16,–

Vol. 368: R. J. Milgram, Unstable Homotopy from the Stable Point of View. V, 109 pages. 1974. DM 16,–

Vol. 369: Victoria Symposium on Nonstandard Analysis. Edited by A. Hurd and P. Loeb. XVIII, 339 pages. 1974. DM 26,–

Vol. 370: B. Mazur and W. Messing, Universal Extensions and One Dimensional Crystalline Cohomology. VII, 134 pages. 1974. DM 16,–

Vol. 371: V. Poenaru, Analyse Différentielle. V, 228 pages. 1974. DM 20,–

Vol. 372: Proceedings of the Second International Conference on the Theory of Groups 1973. Edited by M. F. Newman. VII, 740 pages. 1974. DM 48,–

Vol. 373: A. E. R. Woodcock and T. Poston, A Geometrical Study of the Elementary Catastrophes. V, 257 pages. 1974. DM 22,–

Vol. 374 S. Yamamuro, Differential Calculus in Topological Linear Spaces. IV, 179 pages. 1974. DM 18,–

Vol. 375: Topology Conference 1973. Edited by R. F. Dickman Jr. and P. Fletcher. X, 283 pages. 1974. DM 24,–

Vol. 376: D. B. Osteyee and I. J. Good, Information, Weight of Evidence, the Singularity between Probability Measures and Signal Detection. XI, 156 pages. 1974. DM 16.–

Vol. 377: A. M. Fink, Almost Periodic Differential Equations. VIII, 336 pages. 1974. DM 26,–

Vol. 378: TOPO 72 – General Topology and its Applications. Proceedings 1972. Edited by R. Alò, R. W. Heath and J. Nagata. XIV, 651 pages. 1974. DM 50,–

Vol. 379: A. Badrikian et S. Chevet, Mesures Cylindriques, Espaces de Wiener et Fonctions Aléatoires Gaussiennes. X, 383 pages. 1974. DM 32,–

Vol. 380: M. Petrich, Rings- and Semigroups. VIII, 182 pages. 1974. DM 18,–

Vol. 381: Séminaire de Probabilités VIII. Edité par P. A. Meyer. IX, 354 pages. 1974. DM 32,–

Vol. 382: J. H. van Lint, Combinatorial Theory Seminar Eindhoven University of Technology. VI, 131 pages. 1974. DM 18,–

Vol. 383: Séminaire Bourbaki – vol. 1972/73. Exposés 418-435 IV, 334 pages. 1974. DM 30,–

Vol. 384: Functional Analysis and Applications, Proceedings 1972. Edited by L. Nachbin. V, 270 pages. 1974. DM 22,–

Vol. 385: J. Douglas Jr. and T. Dupont, Collocation Methods for Parabolic Equations in a Single Space Variable (Based on C^1-Piecewise-Polynomial Spaces). V, 147 pages. 1974. DM 16,–

Vol. 386: J. Tits, Buildings of Spherical Type and Finite BN-Pairs. IX, 299 pages. 1974. DM 24,–

Vol. 387: C. P. Bruter, Eléments de la Théorie des Matroïdes. V, 138 pages. 1974. DM 18,–

Vol. 388: R. L. Lipsman, Group Representations. X, 166 pages. 1974. DM 20,–

Vol. 389: M.-A. Knus et M. Ojanguren, Théorie de la Descente et Algèbres d' Azumaya. IV, 163 pages. 1974. DM 20,–

Vol. 390: P. A. Meyer, P. Priouret et F. Spitzer, Ecole d'Eté de Probabilités de Saint-Flour III – 1973. Edité par A. Badrikian et P.-L. Hennequin. VIII, 189 pages. 1974. DM 20,–

Vol. 391: J. Gray, Formal Category Theory: Adjointness for 2-Categories. XII, 282 pages. 1974. DM 24,–

Vol. 392: Géométrie Différentielle, Colloque, Santiago de Compostela, Espagne 1972. Edité par E. Vidal. VI, 225 pages. 1974. DM 20,–

Vol. 393: G. Wassermann, Stability of Unfoldings. IX, 164 pages. 1974. DM 20,–

Vol. 394: W. M. Patterson 3rd, Iterative Methods for the Solution of a Linear Operator Equation in Hilbert Space – A Survey. III, 183 pages. 1974. DM 20,–

Vol. 395: Numerische Behandlung nichtlinearer Integrodifferential- und Differentialgleichungen. Tagung 1973. Herausgegeben von R. Ansorge und W. Törnig. VII, 313 Seiten. 1974. DM 28,–

Vol. 396: K. H. Hofmann, M. Mislove and A. Stralka, The Pontryagin Duality of Compact O-Dimensional Semilattices and its Applications. XVI, 122 pages. 1974. DM 18,–

Vol. 397: T. Yamada, The Schur Subgroup of the Brauer Group. V, 159 pages. 1974. DM 18,–

Vol. 398: Théories de l'Information, Actes des Rencontres de Marseille-Luminy, 1973. Edité par J. Kampé de Fériet et C. Picard. XII, 201 pages. 1974. DM 23,–